W9-BNT-483

THE Mystery Chronicles

More Real-Life X-Files

THE Mystery Chronicles

MORE REAL-LIFE X-FILES

Joe Nickell

THE UNIVERSITY PRESS OF KENTUCKY

Publication of this volume was made possible in part by a grant from the National Endowment for the Humanities.

Scholarly publisher for the Commonwealth, serving Bellarmine University, Berea College, Centre College of Kentucky, Eastern Kentucky University, The Filson Historical Society, Georgetown College, Kentucky Historical Society, Kentucky State University, Morehead State University, Murray State University, Northern Kentucky University, Transylvania University, University of Kentucky, University of Louisville, and Western Kentucky University.

Editorial and Sales Offices: The University Press of Kentucky
663 South Limestone Street, Lexington, Kentucky 40508-4008
www.kentuckypress.com

08 07 06 05 04 5 4 3 2

Library of Congress Cataloging-in-Publication Data

Nickell, Joe.
The mystery chronicles : more real-life X-files / Joe Nickell.
p. cm.
Includes bibliographical references and index.
ISBN 0-8131-2318-6 (alk. paper)
1. Parapsychology—Case studies. 2. Curiosities and wonders.
3. Impostors and imposture. I. Title.
BF1031.N517 2004
001.94—dc22 003020561

This book is printed on acid-free recycled paper meeting the requirements of the American National Standard for Permanence in Paper for Printed Library Materials.

Printed in the United States of America on acid-free paper.

Member of the Association of
American University Presses

In Memory of
STEVE ALLEN
(1921–2000)

Entertainer, television pioneer, actor, writer, skeptic, and friend

Contents

Acknowledgments xi

Foreword xiii

Introduction xv

1 *Mystery of the Nazca Lines* 1

2 *The Fiery Specter* 10

3 *The Exorcist: The Case Behind the Movie* 14

4 *The "Goatsucker" Attack* 28

5 *Undercover Among the Spirits: Investigating Camp Chesterfield* 31

6 *Alien Hybrid?* 46

7 *Image of Guadalupe: Myth-perception* 51

8 *Human Blowtorch* 56

9 *Remotely Viewed? The Charlie Jordan Case* 61

10 *Amityville: The Horror of It All* 73

11 *Sideshow! Investigating Carnival Oddities and Illusions* 78

12 *"Mothman" Solved! Investigating on Site* 93

13 *Relics of the Headless Saint* 100

14 *Circular Reasoning: Crop Circles and Their "Orbs" of Light* 115

15 *Zanzibar Demon* 124

16 *Winchester Mystery House* 128

17 *Voodoo in New Orleans* 140

18 *Secrets of the Voodoo Tomb* 152

19 *A Case of "SHC" Demystified* 162

20 *Tracking the Swamp Monsters* 165

21 *John Edward: Talking to the Dead?* 176

22 *Scandals and Follies of the "Holy Shroud"* 187

23 *"Pyramid Power" in Russia* 200

24 *Diagnosing the "Medical Intuitives"* 207

25 *Alien Abductions as Sleep-Related Phenomena* 218

26 *"Visitations": After-Death Contacts* 228

27 *The Sacred Cloth of Oviedo* 241

28 *A Typical Aries?* 248

29 *The Case of the Psychic Shamus: Do Psychics Really Help Solve Crimes?* 252

30 *The Pagan Stone* 258

31 *Benny Hinn: Healer or Hypnotist?* 261

32 *Australia's Convict Ghosts* 271

33 *Psychic Pets and Pet Psychics* 277

34 *Cryptids "Down Under"* 289

35 *Joseph Smith: A Matter of Visions* 296

36 *In Search of Fisher's Ghost* 304

37 *Ghostly Portents in Moscow* 311

38 *Mystique of the Octagon Houses* 315

39 *Weeping Icons* 324

40 *Spiritualist's Grave* 331

41 *Incredible Stories: Charles Fort and His Followers* 335

INDEX 341

Acknowledgments

I am supremely grateful to John and Mary Frantz for their generous establishment of an investigative fund that helps make much of my research possible.

I am once again indebted to my colleagues at the Center for Inquiry (CFI) in Amherst, New York, where—since 1995—I have been Senior Research Fellow of the Committee for the Scientific Investigation of Claims of the Paranormal (CSICOP). These colleagues include Paul Kurtz, chairman; Barry Karr, executive director; Kevin Christopher, director of public relations; Kendrick Frazier, editor of CSICOP's official magazine, *Skeptical Inquirer*; and Benjamin Radford, managing editor of *Skeptical Inquirer*. All of these helped in numerous ways.

Research assistance was generously provided by Timothy Binga, director of the Center for Inquiry Libraries. Ranjit Sandhu also provided valuable research, and his considerable word-processing skills repeatedly transformed scribbles into finished articles and articles into this book. Colleague Tom Flynn, editor of *Free Inquiry*, often provided expert photographic assistance. So did non-colleague (but good friend) Rob McElroy.

Specific production assistance came from Lisa A. Hutter, art director, and Paul Loynes, production. Taking care of many financial problems were Pat Beauchamp and Paul Paulin.

I continue to be deeply indebted to Robert A. Baker, Emeritus Professor of Psychology, University of Kentucky, and John F. Fischer, forensic analyst (retired) at the Orange County, Florida, Sheriff's Department crime laboratory, who have been invaluable over the years as fellow investigators, coauthors, and friends.

I also continue my indebtedness to fellow CSICOP Executive Council members James E. Alcock, Barry Beyerstein, Thomas Casten, Kendrick Frazier, Martin Gardner, Ray Hyman, Lawrence Jones, Philip J. Klass, Lee Nisbet, Amardeo Sarma, Béla Scheiber, and, again, Paul Kurtz.

In addition to the many individuals mentioned in the text, I am also grateful to the following: George E. Abaunza, Brant Abrahamson, Christian L. Ambrose, Dan Barber, Ed and Diane Buckner, James F. Cherry, Eva Clarke, Christopher Densmore, Christopher Hoolihan, Shozo Kagoshima, Joyce LaJudice, William Loos, Vaughn Rees, Etienne Ríos, Michael Sandras, Mark Scerbo, Marge Sharp, William Sierichs Jr., Glenn Taylor, Jim Underdown, Cynthia Van Ness, Vance Vigrass, and Dana Walpole.

Outside of the United States, I am also very appreciative to the following: From England, Jane Topping (BBC); in Australia, Ian Bryce, Georgina Keep, and Barry Williams; and in Moscow, Kirill Boliakiu.

This is only the short list. To the many other associates at the Center for Inquiry, my friends and fellow skeptics around the world, and the many others who helped in some way, I express my sincere appreciation.

Foreword

I've known Dr. Joe Nickell since he was a young man. That was when he was a budding conjuror, eagerly soaking up every secret and angle of the deception-for-entertainment trade. I could hardly have guessed that after entering the conjuring business he would then change direction to become a professional sleuth, and turn out a stream of fascinating books dealing with so many aspects of human frailty and the vultures who await those who have weakened and become potential fare. *The Mystery Chronicles* is his latest assault against nonsense.

Well equipped by both experience and academic background, this author has literally roamed the globe sniffing for—and finding!—his quarry. From the Shroud of Turin to document forgeries, whether in Canada or in Peru, no fakery is immune to his probing. He'll track a furry monster or poke at a holy icon, with the same strong interest and determination. Unlike so many other "experts," he doesn't sit back thumbing through the writings of others to glean enough information for a book; Joe actually gets out there in the field and runs down the data; he's physically there, pencil and lens poised and ready. His energy is surpassed only by his enthusiasm and dedication to his trade.

In *The Mystery Chronicles*, author Nickell has spanned a remarkable spectrum. It's hard to think of a scoundrel he hasn't skewered, to discover a stone he hasn't turned over, or find a tome he hasn't sifted through to develop his subjects, in this book. In scholarly fashion, he lays out his case, provides the references and the evidence, makes his argument, gives his conclusions, and slams the door on the pretensions and flummery he has exposed to the withering light of day.

We'd be poorer for his absence, and we are served well by his efforts. *The Mystery Chronicles* is an exciting page-turner, just another salvo in the ongoing battle, but it's a loud one, and it's right on target. The contents could bring us to a degree of frustration and anger—did we not know that Joe Nickell still has a few more books in his head, ready to join the battle against deception, misinformation, fakery, and the swindlers who continue to assail us with these weapons. They arise every minute to confront us.

Truth, as Joe knows, is their greatest enemy. . .

James Randi
President of the James Randi Educational Foundation

Introduction

For more than three decades, I have been investigating paranormal claims—that is, those supposedly beyond the range of science and normal human experience. My first important case transpired in 1972 when I investigated Canada's most famous "haunted" place: Mackenzie House in downtown Toronto. As it turned out, I easily found plausible explanations for the various reported phenomena. For example, the sounds of heavy footsteps on the stairs came from an iron staircase in the building next door—just 40 inches away.

At that time I was working as a professional stage magician and "mentalist." I soon went on to become a private investigator, working—mostly undercover—to solve grand theft and other crimes. I sought out that "role" in part to gain skills I knew I could use in my avocation as that other kind of PI: paranormal investigator.

Many years (and many other roles and investigations) later, I returned to the University of Kentucky to earn a master's degree (1982) and doctorate (1987) in English literature, including folklore. I taught there until mid-1995, when I moved to Buffalo to become Senior Research Fellow of the Committee for the Scientific Investigation of Claims of the Paranormal (CSICOP), publisher of *Skeptical Inquirer* magazine. I thus began to investigate strange mysteries full time and to report my findings in a column, "Investigative Files."

Over the years I have learned that, with regard to paranormal claims, people tend to divide into opposing camps, typically styling themselves either as "believers" or as "debunkers." These polarized groups often have more in common than they would admit: primary among these commonalities is the tendency to start with an answer and work backward to the evidence, picking and choosing the "facts" that support their convictions.

I disparage both too-credulous and too-dismissive attitudes, holding that mysteries should actually be investigated in an attempt to solve them. That is the approach of this casebook, which reports some of my most intriguing investigations.

I have examined countless claims and sites, not only across the United States, but around the world. Examples include a case of stigmata in Canada; supernatural relics of a headless saint in Spain; crop circles in England; a weeping icon and "pyramid power" in Russia; a famous ghost tale and the legendary "yowie" (a Bigfoot-type creature) in Australia; and an "alien hybrid" in Germany—among many others.

Over the years I have spent nights in allegedly haunted places, interviewed "alien abductees," exposed phony psychics, gone undercover (and in disguise) to reveal spiritualist mediums' tricks, recreated one of the giant Nazca ground drawings of Perú, and engaged in many similar hands-on investigative activities.

With this collection, I invite readers to come along with me in pursuit of some of these real-life mysteries. I am mindful of Sherlock Holmes's admission (in *A Study in Scarlet*): "A conjurer gets no credit when once he has explained his trick; and if I show you too much of my method of working, you will come to the conclusion that I am a very ordinary individual after all." But come along anyway and we shall see the truth of another of Holmes's remarks (in *The Red-headed League*): "For strange effects and extraordinary combinations we must go to life itself, which is always far more daring than any effort of the imagination."

THE Mystery Chronicles

More Real-Life X-Files

1

Mystery of the Nazca Lines

Etched across 30 miles of gravel-covered desert near Perú's southern coast are the famous Nazca lines and giant ground drawings.

This huge sketchpad was brought to public prominence by Erich von Däniken's *Chariots of the Gods?*—a book that consistently underestimates the abilities of ancient "primitive" peoples and assigns many of their works to visiting extraterrestrials. Von Däniken (1970) argues that the Nazca lines and figures could have been "built according to instructions from an aircraft." He adds: "Classical archaeology does not admit that the pre-Inca peoples could have had a perfect surveying technique. And the theory that aircraft could have existed in antiquity is sheer humbug to them."

Von Däniken does not consider it humbug, and he obviously envisions flying saucers hovering above and beaming down instructions for the markings to awed primitives (presumably in their native tongue).

He views the large drawings as "signals" (von Däniken 1970) and the longer and wider of the lines as "landing strips" (von Däniken 1972). But would extraterrestrials create signals for themselves in the shape of spiders and monkeys? And would such "signals" be less than 80 feet long (like some of the smaller Nazca figures)?

As to the landing-strip notion, Maria Reiche, the German-born mathematician who for years has mapped and attempted to preserve the markings, had a ready rejoinder. Noting that the imagined runways are clear of stones and that the underlying ground is quite soft, she said, "I'm afraid the spacemen would have gotten stuck" (McIntyre 1975).

It is difficult to take von Däniken seriously, especially since his "theory" is not his own and was originated in jest. Paul Kosok (1947), the first to study the markings, wrote: "When first viewed from the air, [the lines] were nicknamed prehistoric landing fields and jokingly compared with the so-called canals on Mars." Moreover, one cropped photo exhibited by von Däniken (1970), showing an odd configuration "very reminiscent of the aircraft parking areas in a modern airport," is actually of the knee joint of one of the bird figures (Woodman 1977). (See Figure 1-1.) Any spacecraft that parked there would have had to be tiny indeed.

Closer to earth, but still merely a flight of fancy, in my opinion, is the notion of Jim Woodman (1977) and some of his colleagues from the International Explorers Society that the ancient Nazcas constructed hot-air balloons for "ceremonial flights," from which they could "appreciate the great ground drawings on the *pampas*." Even if one believes that this theory is also inflated with hot air, one must at least give Woodman credit for the strength of his convictions. Using cloth, rope, and reeds, Woodman and his associates actually made a balloon and gondola similar to those the Nazcas might have made had they actually done so. Woodman and British balloonist Julian Nott then risked their lives in a 300-foot-high flyover of the Nazca plain. When their balloon began descending rapidly, they threw off more and more sacks of bal-

FIGURE 1-1. **Etched upon the Nazca plains in Perú are giant drawings like these. Their large size has fueled misguided speculation that they were drawn with the aid of "ancient astronauts" or by sophisticated surveying techniques, the secrets of which are lost.**

last, but finally had to jump clear of their craft some 10 feet above the pampas. Free of the balloonists' weight, the balloon shot skyward and soared almost out of sight, only to finally crash and drag briefly across the ground.

The Nazca markings are indeed a mystery, although we do know who produced them—von Däniken notwithstanding. Conceding that Nazca pottery is found in association with the lines, von Däniken (1970) writes: "But it is surely oversimplifying things to attribute the geometrically arranged lines to the Nazca culture for that reason alone."

No knowledgeable person does. The striking similarity of the stylized line figures to those of known Nazca art has been clearly demonstrated (Isbell 1978, 1980). In addition to this iconographic evidence must be added that from carbon-14 analysis: Wooden stakes mark the termination of some of the long lines and one of these was dated to C.E. 525 (±80). This is consistent with the presence of the Nazca Indians who flourished in the area from 200 B.C.E. to about C.E. 600. Their graves and the ruins of their settlements lie near the drawings.

The questions of who and when aside, the mystery of *why* the markings were made remains, although several hypotheses have been proffered. One is that they represent some form of offerings to the Indian gods (McIntyre 1975). Another is that they form a giant astronomical calendar or "star chart." Writing in *Scientific American*, William H. Isbell (1978) suggested that an important function of the markings was economic, "related to the drafting of community labor for public works," although at best that is only a partial explanation.

Still another suggestion (first mentioned by Kosok) came from art historian Alan Sawyer (McIntyre 1975): "Most figures are composed of a single line that never crosses itself, perhaps the path of a ritual maze. If so, when the Nazcas walked the line, they could have felt they were absorbing the essence of whatever the drawing symbolized." Sawyer is correct in observing that most of the figures are drawn with a continuous, uninterrupted line. But there *are* exceptions, and it is possible that the continuous-line technique is related to the method of producing the figures, as we shall discuss presently.

In 1991 anthropologist and professor of astronomy Anthony F. Aveni and anthropologist Helaine Silverman reported the results of extensive studies of the Nazca lines. Sightings along some 700 radiating lines showed an unfortunately near randomness with regard to astronomical significance. Rather, the lines correlated with geographic features, suggesting to the researchers that the Nazca line makers were driven by "an inescapable concern about water," and that the Nazcas may have walked or danced along the lines as part of "irrigation ceremonies."

In any case, whatever meaning(s) we ascribe to the Nazca lines and drawings must be considered in light of other giant ground markings elsewhere. In South America, giant effigies are found in other locales in Perú, for example, and in Chile, in the Atacama Desert (Welfare and Fairley 1980). Interestingly, the plan of the Incan city of Cuzco was laid out in the shape of a puma, and its inhabitants were known as "members of the body of the puma" (Isbell 1978, 1980).

Turning to North America, there is the Great Serpent Mound in Ohio and giant effigies in the American Southwest. In 1978, with the aid of an Indian guide, I was able to view the ground drawings near Blythe, California, in the Mojave Desert. However, although they are thought to date from a much later period (Setzler 1952), none of the Blythe figures match the size of the largest Nazca drawings; also, the human figures and horselike creatures are much cruder in form, typically having solid-area bodies and sticklike appendages. Moreover, absent from the Blythe site are the "ruler-straight" lines that may or may not have calendrical significance.

In short, there are both similarities and dissimilarities between the Nazca and other ground drawings that complicate our attempts to explain them. Certainly the Blythe and other effigies have no attendant von Dänikenesque "runways"; neither do their crude forms suggest that they were drawn with the aid of hovering spacecraft.

It seemed to me that a study of *how* the lines were planned and executed might shed some light on the ancient riddle. English explorer and filmmaker Tony Morrison has demonstrated that, by using a series of ranging poles, straight lines can be constructed over many miles (Welfare and Fairley 1980). (The long lines "veer from a straight line by only a few yards every mile," reports *Time* [Mystery 1974].) In fact, along some lines, the remains of posts have been found at roughly one-mile intervals (McIntyre 1975).

By far the most work on the problem of Nazca engineering methods was done by Maria Reiche (1976). She explained that Nazca artists prepared preliminary drawings on small six-foot-square plots. These plots are still visible near many of the larger figures. The preliminary drawing was then broken down into its component parts for enlargement. Straight lines, she observed, could be made by stretching a rope between two stakes. Circles could easily be scribed by means of a rope anchored to a rock or stake, and more complex curves could be drawn by linking appropriate arcs. As proof, she reported that there are indeed stones or holes at points that are centers for arcs.

Reiche did not, however, detail the specific means for positioning the stakes that apparently served as the centers for arcs or the endpoints of straight lines. In her book she wrote, "Ancient Peruvians must have had instruments and equipment which we ignore and which together with ancient knowledge were buried and hidden from the eyes of the conquerors as the one treasure which was not to be surrendered."

Isbell (1978) suggested that the Nazcas used a grid system adapted from their weaving experience, a loom "establishing a natural grid within which a figure is placed." All that would be necessary, he observed, would be simply to enlarge the grid to produce the large drawings.

However, as one who has used the grid system countless times (in reproducing large trademarks and pictorials on billboards—summer work during my high-school and college years), I am convinced the grid system was *not* employed. To mention only one reason, a characteristic of the grid method is that errors or distortions are largely confined to individual squares. Thus, the "condor" drawing in Figure 1-1—with its askew wings, mismatched feet, and other asymmetrical features—seems not to have been produced by means of a grid.

Other, even less likely possibilities are the plotting of points by a traverse surveying technique (such as is used today to plot a boundary of land) or by triangulation. Having some experience with both of these, I note that such methods depend on the accurate measurement of angles, and there appears to be no evidence that the Nazcas had such a capability.

I decided to attempt to reproduce one of the large Nazca figures—the 440-foot-long condor in the center of Figure 1-1—using a means I thought the Nazcas might actually have employed. I was joined in the project by two of my cousins, John May and Sid Haney. The method we chose was quite simple: We would establish a center line and locate points on the ground drawing by plotting their coordinates. That is, on the small drawing we would measure along the center line from one end (the bird's beak) to a point on the line directly opposite the point to be plotted (say, a wing tip). Then we would measure the distance on the ground from the center line to the desired point. A given number of units on the small drawing would require the same number of units—larger ones—on the large drawing. We used only a set of sticks, set at right angles, for sighting, and two lengths of marked and knotted cord for measuring.

My father, J. Wendell Nickell, took charge of logistics; my young cousin Jim Mathis and nephew Con Nickell completed our crew. After we had marked the figure—a total of less than two days' work—pilot Jerry Mays flew us over the drawing at just under 1,000 feet. John took photos while leaning out of the banked plane, with me holding onto his belt. (For more details of the construction, see Nickell 1983.) *Scientific American* magazine (Big Picture 1983) later termed our production "remarkable in its exactness" to the Nazca original (see Figures 1-2 and 1-3).

In summary, we do know that it was the Nazca peoples who produced the drawings. Although the large size of the drawings does suggest the possibility that they were meant to be viewed from above, as by the Indian gods, the figures can be recognized, at least to some extent, from the ground. The drawings could have been produced by a simple method requiring only materials available to South American Indians

FIGURE 1-2. The author's duplication of the giant "condor" drawing, made full size and using only sticks and cord such as the Nazcas might have employed. The experimental drawing—possibly the world's largest art reproduction—is viewed here from just under 1,000 feet.

FIGURE 1-3. Detail showing the author standing in one of the giant bird's claws.

centuries ago. The Nazcas probably used a simplified form of this method, with perhaps a significant amount of the work done freehand. There is no evidence that extraterrestrials were involved; but if they were, one can only conclude that they seem to have used sticks and cord just as the Indians did.

REFERENCES

Aveni, Anthony F., and Helaine Silverman. 1991. Between the lines: Reading the Nazca markings as rituals writ large. *The Sciences* (July/August): 36–42.

Isbell, William H. 1978. The prehistoric ground drawings of Perú. *Scientific American* 239 (October): 140–53.

———. 1980. Solving the mystery of Nazca. *Fate* (October): 36–48.

Kosok, Paul (with Maria Reiche). 1947. The markings of Nazca. *Natural History* 56: 200–38.

McIntyre, Loren. 1975. Mystery of the ancient Nazca lines. *National Geographic* (May): 716–28.

Mystery on the mesa. 1974. *Time*, March 25.

Nickell, Joe. 1983. The Nazca drawings revisited: Creation of a full-size duplicate. *Skeptical Inquirer* 7, no. 3 (Spring): 36–44.

Reiche, Maria. [1968] 1976 (rev. ed.). *Mystery on the Desert* . Stuttgart: Privately printed.

Setzler, Frank M. 1952. Seeking the secrets of the giants. *National Geographic* 102: 393–404.

The big picture. 1983. *Scientific American* June, 84.

Von Däniken, Erich. 1970. *Chariots of the Gods?* New York: G. P. Putnam.

———. 1972. *Gods from Outer Space*. New York: Bantam Books.

Welfare, Simon, and John Fairley. 1980. *Arthur C. Clarke's Mysterious World*. New York: A & W Publishers.

Woodman, Jim. 1977. *Nazca: Journey to the Sun*. New York: Pocket Books.

2

The Fiery Specter

Gruesome deaths attributed to spontaneous human combustion (SHC) continue to intrigue the public. Arch-promoter of SHC, Larry E. Arnold, has produced a weighty tome on the subject, entitled *Ablaze! The Mysterious Fires of Spontaneous Human Combustion* (1995). Unfortunately, the cases Arnold hypes have very plausible—indeed, probable—explanations. For example, in the 1951 case of Mary Reeser in St. Petersburg, Florida, it was established that when last seen she was wearing flammable nightclothes and smoking a cigarette, after having taken Seconal sleeping pills. The extreme destruction of her body was almost certainly due to the "wick effect," in which the body's melted fat is retained by clothing, carpeting, chair stuffing, and so on; this fuels still more fire to cause still more destruction (Nickell 2001).

Case after case put forward by Arnold and other mystery mongers succumbs to investigation. One such case in *Ablaze!* is that of a "baffling" and "abnormal fiery accident" that occurred "about fifteen miles southeast of Baltimore, in Arundel [*sic*] County, Maryland" (actually Anne Arundel County). The date is rather vaguely given as "early April 1953"—a curious way of expressing it, as the accident transpired on April 1 ("Man hurt" 1953). Arnold provides not a single source citation for the case other than a brief quotation from the late Frank Edwards, one-time columnist for *Fate* magazine and author of several mystery-mongering books, like *Stranger Than Science*, notorious for their errors and exaggerations.

As Arnold relates the case:

> Here, Maryland and State Police found Bernard J. Hess in his overturned car at the bottom of a twenty-foot embankment. The Baltimore man had a fractured skull. Therefore, the cause of death appeared obvious, the case routine. Then the coroner investigated. Routine quickly ceased. Although he found no trace of fire damage to the wreckage, the coroner discovered first- and second-degree burns covered two-thirds of the dead man's fully clothed body. Police failed initially to notice Hess's searing because . . . well, *because his garments hadn't burned!*
>
> Authorities concluded that Hess's severely blistered skin would make it impossibly painfully [*sic*] for him to dress himself after being burned. Contemporary reports do not mention officials finding any electrical or fuel problems with the car that would have caused his injuries.

Arnold continues:

> Did Hess fall victim to foul play at the scene, as unknown assailants stripped Hess naked, doused him with unidentified chemical accelerants and lit them, then

> re-dressed and drove their victim to a location remote from the crime to push him over the embankment in his car? No evidence supported this. Did Hess succumb to SHC?

As Frank Edwards remarked three years after this incident: "The burns which played a part in his death constitute another mystery which remains unsolved."

Whereas Arnold insists that "contemporary reports" give no clue to the mystery, in fact newspaper accounts actually report the medical examiner's official determination. First, however, the reader is invited to provide a plausible solution to the mystery. There are several potential hypotheses, each more credible than spontaneous human combustion, but you may ignore Arnold's deliberately silly scenario of "unknown assailants" stripping, burning, and re-dressing the victim. Instead, simply consider the circumstances of an overturned car and the damaged skin coupled with unburned clothing. Please pause here to construct your hypothesis.

Finished? My own analysis began with the possibility that Mr. Hess had simply been scalded by hot water from a ruptured radiator—or from the heater core located in the dashboard. Apparently that was not the actual mechanism, but it certainly represents a hypothetical solution. According to the *Baltimore Sun*, "Gasoline 'burns' on the body of Bernard Joseph Hess . . . ha[d] nothing to do with his death, an autopsy yesterday disclosed" ("'Burns'" 1953). Dr. Russell S. Fisher, Baltimore's chief medical examiner, stated that the 35-year-old Hess died of head injuries suffered when the convertible he was driving overturned on April 1. Dr. Fisher said that gasoline had soaked through the victim's clothing to inflict what the *Sun* called "skin injuries similar to burns," caused by a reaction to the fuel. The *Baltimore News-Post* ("Mystery burns" 1953) cited an assistant medical examiner who provided a concurring opinion: "The examiner, Dr. Francis J. Januszeski, said gasoline is an 'organic solvent,' used in cleaning to remove grease, and has somewhat the same effect on flesh."

Interestingly, Bernard "Whitey" Hess was a convicted forger who had been released on probation. He had used another man's credentials to pose as a potential auto buyer and thus steal the convertible in which he died. His wife—then serving a sentence for embezzlement—was notified in jail of his death ("Mysterious death" 1953; "'Burns'" 1953).

Obviously, the Hess case had nothing to do with spontaneous human combustion, as Larry Arnold should have realized. Arnold, who is not a physicist but a Pennsylvania school-bus driver, had no justification for asking ominously, "Did Hess succumb to SHC?" The unburned clothing should have led any sensible investigator to one of the possibilities limited by that fact: for example, that Hess had been burned previously, or his skin injuries were caused by steam or hot water, chemical liquids or vapors, or some type of radiation (possibly even extreme sunburn through loosely woven clothing). In any event, Arnold could have done as I did and sought out the newspaper accounts of the day. It would have saved him from yet another folly.

REFERENCES

Arnold, Larry. 1995. *Ablaze! The Mysterious Fires of Spontaneous Human Combustion*. New York: M. Evans and Co., 182–83.

"Burns" on the body of forger ruled out as cause of death. 1953. *Baltimore Sun*, 3 April.

Man hurt in auto crash dies here. 1953. *Baltimore Sun*, 2 April.

Mysterious death of forger probed. 1953. *Baltimore Sun*, 2 April, evening edition.

Mystery burns caused by gasoline, doctor reports. 1953. *Baltimore News-Post*, 2 April.

Nickell, Joe. 2001. *Real-Life X-Files*. Lexington, Ky.: University Press of Kentucky, 28–36.

3

The Exorcist

The Case Behind the Movie

Belief in demonic possession is getting a new propaganda boost. Not only has the 1973 horror movie *The Exorcist* been re-released, but the "true story" that inspired it is also chronicled in a reissued book and a made-for-TV movie, both titled *Possessed* (Allen 2000). However, a year-long investigation by a Maryland writer (Opsasnik 2000), together with my own analysis of events chronicled in the exorcising priest's diary, belie the claim that a teenage boy was possessed by Satan in 1949.

Psychology versus Possession

Belief in spirit possession flourishes in times and places where there is ignorance about mental states. Citing biblical examples, the medieval Church taught that demons were able to take control of an individual; by the sixteenth century, so-called demonic behavior had become relatively stereotypical. It manifested itself in convulsions, prodigious strength, insensitivity to pain, temporary blindness or deafness, clairvoyance, and other abnormal characteristics. Some early notions of possession may have been based on the symptoms of three brain disorders: epilepsy, migraine, and Tourette's syndrome (Beyerstein 1988). Psychiatric historians have long attributed demonic manifestations to such aberrant mental conditions as schizophrenia and hysteria, noting that as mental illness began to be recognized as such after the seventeenth century, there was a consequent decline in demonic superstitions and reports of possession (Baker 1992, 192). In 1999, the Vatican did update its 1614 guidelines for expelling demons, urging exorcists to avoid mistaking psychiatric illness for possession ("Vatican" 1999).

In many cases, however, supposed demonic possession can be a learned role that fulfills certain important functions for those claiming it. In his book *Hidden Memories: Voices and Visions from Within*, psychologist Robert A. Baker (1992) noted that possession was sometimes feigned by nuns to act out sexual frustrations, protest restrictions, escape unpleasant duties, attract attention and sympathy, and fulfill other psychologically useful or necessary functions.

Many devout claimants of stigmata, inedia, and other powers have also exhibited alleged demonic possession. For example, at Loudon, France, a prioress, Sister Jeanne des Anges (1602–1665), was part of a contagious outbreak of writhing, convulsing nuns. Jeanne herself exhibited stigmatic designs and lettering on her skin. A bloody cross "appeared" on her forehead, and the names of Jesus, Mary, and others were found on her hand—always clustered on her left hand, just as one would expect if a right-handed person were marking them. She went on tour as a "walking relic" and was exhibited in Paris to credulous thousands. There were a few skeptics, but Cardinal Richelieu rejected the idea of testing Jeanne by enclosing her hand in a sealed glove. He felt that such an experiment would amount to testing God (Nickell 1998,

230–31). Interestingly enough, while I was researching and writing this chapter I was called to southern Ontario on a case of dubious possession that also involved stigmata.

Possession can be childishly simple to fake. For example, an exorcism broadcast by ABC's *20/20* in 1991 featured a 16-year-old girl who, her family claimed, was possessed by 10 separate demonic entities. However, to skeptics her alleged possession seemed to be indistinguishable from poor acting. She even stole glances at the camera before affecting convulsions and other "demonic" behavior (Nickell 1998).

Of course, a person with a strong impulse to feign diabolic possession may indeed be mentally disturbed. Although the teenager in the *20/20* episode reportedly improved after the exorcism, it was also pointed out that she continued "on medication" ("The Exorcism" 1991). To add to the complexity, the revised Vatican guidelines also appropriately urge against believing that a person is possessed rather than counting him or her as merely "the victim of [his or her] own imagination" ("Vatican" 1999).

With less modern enlightenment, however, the guidelines also reflect Pope John Paul II's efforts to convince doubters that the devil actually exists. In various homilies John Paul II has denounced Satan as a "cosmic liar and murderer." A Vatican official who presented the revised rite stated, "The existence of the devil isn't an opinion, something to take or leave as you wish. Anyone who says he doesn't exist wouldn't have the fullness of the Catholic faith" ("Vatican" 1999).

Unchallenged by the new exorcism guidelines is the acceptance of such alleged signs of possession as demonstrating supernormal physical force and speaking in unknown tongues. In the case broadcast by *20/20*, the teenage girl did exhibit "tongues" (known as *glossolalia* [Nickell 1998, 103–09]), but it was unimpressive: she merely chanted, "Sanka dali. Booga, booga." She did struggle against the restraining clerics, one of whom claimed that, had she not been held down, *she would have been* levitating! At that point a group of magicians, psychologists, and other skeptics with whom I was watching the video gleefully encouraged, "Let her go! Let her go!" (Nickell 1995).

"True Story"

Demonstrating prodigious strength, speaking in an unknown language, and exhibiting other allegedly diabolical feats supposedly characterize the "true story" behind *The Exorcist*. The 1973 horror movie (starring Linda Blair as the devil-plagued victim) was based on the 1971 bestselling novel of that title by William Peter Blatty. The movie, reported one writer, "somehow reached deep into the subconscious and stirred up nameless fears." Some moviegoers vomited or fainted; others left trembling, and there were "so many outbreaks of hysteria that, at some theaters, nurses and ambulances were on call." Indeed, "[m]any sought therapy to rid themselves of fears they could not explain. Psychiatrists were writing about cases of 'cinematic neurosis'" (Allen 2000, viii–ix).

Blatty had heard about the exorcism performed in 1949 and, almost two decades later, wrote to the exorcist to inquire about it. However, the priest, Father William S. Bowdern, declined to assist Blatty because he had been directed by the archbishop to keep the matter secret. He did tell Blatty—then a student at Washington's Georgetown University, a Jesuit institution—about the diary an assisting priest had kept of the disturbing events (Allen 2000, ix–x).

The diary, written by Father Raymond J. Bishop, consisted of an original 26-page, single-spaced typescript and three carbon copies. One of the copies was eventually provided to Thomas B. Allen, author of *Possessed*, and included as an appendix to the 2000 edition of Allen's book. The copy came from Father Walter Halloran, who had also assisted with the exorcism. Halloran verified the authenticity of the diary and stated that it had been read and approved by Bowdern (Allen 2000, 243, 301).

The diary opens with a "Background of the Case." The boy, an only child identified as "R," was born in 1935 and raised an Evangelical

Lutheran like his mother. His father was baptized a Catholic but had had "no instruction or practice" in the faith. The family's Cottage City, Maryland, home included the maternal grandmother, who had been a "practicing Catholic until the age of fourteen years" (Bishop 1949, 245).

On January 15, 1949, R and his grandmother heard odd "dripping" and scratching noises in her bedroom, where a picture of Jesus shook "as if the wall back of it had been bumped." The effects lasted 10 days but were attributed to a rodent. Then R began to say he could hear the scratching when others could not. Soon a noise like that of "squeaking shoes" (or, one wonders, could it have been bedsprings?) became audible, though it "was heard only at night when the boy went to bed." On the sixth evening, the scratching noise resumed, and R's mother and grandmother lay with him on his bed, whereupon they "heard something coming toward them similar to the rhythm of marching feet and the beat of drums." The sound seemed to "travel the length of the mattress and back again" repeatedly (Bishop 1949, 246). Was R tapping his toes against the bed's footboard?

Poltergeists and Ouija Spirits

At this point the case was exhibiting features often attributed to a *poltergeist* (or "noisy spirit"). Poltergeist phenomena typically involve disturbances—noises, movement of objects, or, rarely, serious effects like outbreaks of fire—typically centering around a disturbed person, usually a child. Believers often attribute the occurrences to "psychokinetic energy" or other mystical force imagined to be produced by the repressed hostilities of the pubescent child. Skeptics can agree with all but the mystical part, observing that one unknown cannot be explained by invoking another. Skeptics have a simpler explanation, attributing the effects to the cunning tricks of a naughty youth or occasionally a disturbed adult. When such cases have been properly investigated—by magicians and detectives using hidden cameras, lie detectors, tracer powders (dusted on objects likely to be involved), and other techniques—they usually turn out to be the pranks of young or immature mischief-makers.

Consider some of the "other manifestations" associated with R in the early part of the case, as recorded in the diary:

> An orange and a pear flew across the entire room where R was standing. The kitchen table was upset without any movement on the part of R. Milk and food were thrown off the table and stove. The bread-board was thrown on to the floor. Outside the kitchen a coat on its hanger flew across the room; a comb flew violently through the air and extinguished blessed candles; a Bible was thrown directly at the feet of R, but did not injure him in any way. While the family was visiting a friend in Boonesboro, Maryland, the rocker in which R was seated spun completely around through no effort on the part of the boy. R's desk at school moved about on the floor similar to the plate on a Ouija board. R did not continue his attendance out of embarrassment [Bishop 1949, 248].

It is well to consider here the sage advice of the late Milbourne Christopher, investigator and magician, not to accept statements of what actually happened from the suspected "poltergeist." Regarding one such case, Christopher (1970, 149–60) pointed out that all that was necessary to see the events not as paranormal occurrences but as deliberate deceptions was to "suppose that what the boy said was not true, that he was in one room when he said he was in another in some instances. Also let us suppose that what people thought they saw and what actually happened were not precisely the same." Experience shows that even "reliable witnesses" are capable of being deceived. As one confessed "poltergeist"—an 11-year-old girl—observed: "I didn't throw all those things. People just imagined some of them" (Christopher 1970, 149). In the case of R, we must realize that the previously described events (the flying fruit, etc.) were not witnessed by Father Bishop; he reported them in his diary as background to the case, and so the reports were necessarily second-hand or worse.

In fact, it was the trickery behind the poltergeist-like disturbances of 1848 that launched modern spiritualism. As the Fox sisters confessed decades later, their pretended spirit contact began as the pranks of "very mischievous children" who began their shenanigans "to terrify our dear mother, who was a very good woman and very easily frightened" and who "did not suspect us of being capable of a trick because we were so young." As Margaret Fox explained, the schoolgirls threw slippers at a disliked brother-in-law, shook the dinner table, and produced noises by bumping the floor with an apple on a string and by knocking on the bedstead (Nickell 1995).

The Fox sisters were followed in 1854 by the Davenport brothers, schoolboys Ira and William, who were the focus of odd events that included cutlery dancing about the family's kitchen table. Ira sometimes claimed that, when alone, spirits had whisked him to distant spots. Soon the boys advanced to spirit-rapped messages, "trance" writing and speaking, and other "spirit manifestations." In his old age, Ira confessed to magician/paranormal investigator Harry Houdini that the brothers' spirit communication—which launched and maintained their careers as two of the world's best-known spiritualistic mediums—had all been produced by trickery. Indeed, they had been caught in deceptions many times (Nickell 1999). (For more on the Davenports, see chapter 40, "Spiritualist's Grave.")

The Foxes and Davenports are not isolated examples. It should therefore not be surprising to learn that the case of R, which began as a seeming poltergeist outbreak, soon advanced to one of alleged spirit communication, before finally escalating to one of supposed diabolic possession.

R had been close to an aunt, who often visited from St. Louis. A devoted Spiritualist, she introduced R to the Ouija board. With their fingers on the planchette, they saw it move about the board's array of printed letters, numbers, and the words *yes* and *no* to spell out messages, which she told R were from spirits of the dead. (Actually, as skeptics know, the planchette is moved not by spirits but by the sitters' involuntary—or voluntary!—muscular control [Nickell 1995, 58].) This aunt also told R and his mother how, "lacking a Ouija board, spirits could try to get through to this world by rapping on walls" (Allen 2000, 2).

R had played with the Ouija board by himself. Then began the outbreak of noises, and 11 days later he was devastated by news of his aunt's death in St. Louis. He returned to the Ouija board, spending hours at the practice, and "almost certainly" used it to try to contact his beloved aunt (Allen 2000, 2–6). As R, his mother, and grandmother lay in R's bed and listened to the drumming sound, his mother wondered aloud whether this was the aunt's spirit. If so, she added, "Knock three times" (thus adopting a practice of the Fox sisters). Thereupon, the diary records that the three felt "waves of air" striking them and heard distinct knocks followed by "claw scratchings on the mattress."

Possession?

For approximately four continuous nights thereafter, markings appeared on the teenager's body, after which the clawlike scratches took the form of printed words. Whenever the scratching noise was ignored, the mattress began to shake, at times violently, and at one time the coverlet was pulled loose (Bishop 1949, 246–47).

R's parents were becoming frantic. They had watched their son become unruly, even threatening to run away, and he seemed to be "on the verge of violence" (Allen 2000, 57). They sought help from a physician, who merely found the boy "somewhat high-strung," and then from a psychologist, whose opinions went unrecorded. A psychiatrist found R to be "normal," but "declared that he did not believe the phenomena." A Spiritualist and two Lutheran ministers were consulted (Bishop 1949, 248). One of the latter eventually advised the parents, "You have to see a Catholic priest. The Catholics know about things like this" (Allen 2000, 24).

A young priest was called in, but the boy's condition was worsening and R was admitted to a Jesuit hospital some time between February 27 and March 6. The priest, Father E. Albert Hughes, prepared for an exorcism as the alleged poltergeist and demonic outbreaks and manifestations intensified. Reportedly, the nuns "couldn't keep the bed still," scratches appeared on R's chest, and he began to curse in "a strange language." A later source said it was Aramaic, but a still later "well-documented record" failed to mention "any such language competence" (Allen 2000, 36). The attempted exorcism reportedly ended abruptly when the boy, who had slipped a hand free and worked loose a piece of bedspring, slashed Hughes's arm from shoulder to wrist, inflicting a wound that took more than 100 stitches (Allen 2000, 37).

One investigator, however, doubts whether this attack—or even this first exorcism—ever occurred, having searched in vain for corroborative evidence (Opsasnik 2000). In any event, the parents considered making a temporary move to St. Louis, where relatives lived. When this possibility was discussed, the word "Louis" appeared across R's ribs; when the question arose as to when, "Saturday" was seen plainly on his hip; and when the duration was considered, "3½ weeks" appeared on his chest. The possibility that R was producing the markings was dismissed on the ground that his mother "was keeping him under close supervision," but they might have been made previously and only revealed as appropriate, or he might have produced them as he feigned being "doubled up" and screaming in pain.

According to the diary, "The markings could not have been done by the boy for the added reason that on one occasion there was writing on his back" (Bishop 1949, 247). Such naïve thinking is the reason "poltergeists" are able to thrive. A determined youth, probably even without a wall mirror, could easily have managed such a feat—if it actually occurred. Although the scratched messages proliferated, they never again appeared on a difficult-to-reach portion of the boy's anatomy.

In St. Louis, more poltergeist-type effects were manifested, whereupon Father Bishop (the diarist) was drawn to the case. Bishop left a bottle of holy water in R's bedroom, but later—while the boy claimed to have been dozing—it went sailing across the room. On another occasion, R's parents found the way into his room blocked by a 50-pound bookcase. A stool "fell over." Initially, Bishop and another priest, Father William Bowdern, believed that R could have deliberately produced all of the phenomena that had occurred thus far in St. Louis; they apparently recognized that stories of alleged incidents in Maryland were merely interesting hearsay (Allen 2000, 61–76).

Eventually Bowdern changed his view and was instructed by Archbishop Joseph Ritter to perform an exorcism on R. Bowdern was accompanied by Father Bishop and Walter Halloran (mentioned earlier as providing a copy of the diary to author Allen), who was then a Jesuit student. Bowdern began the ritual of exorcism in R's room. Scratches began to appear on the boy's body, including the word "HELL" on his chest "in such a way that R could look down upon his chest and read the letters plainly." A "picture of the devil" also appeared on the boy's leg. "Evidently the exorcism prayers had stirred up the devil," the diary states, because, after a period of sleep R "began sparring" and "punching the pillow with more than ordinary force" (Bishop 1949, 255–57).

Soon Bowdern "believed deep in his soul that he was in combat with Satan" (Allen 2000, 117). R thrashed wildly; he spat in the faces of the priests and even his mother; he contorted and lashed out; he urinated. Reports the diary:

> From 12:00 midnight on, it was necessary to hold R during his fights with the spirit. Two men were necessary to pin him down to the bed. R shouted threats of violence at

> them, but vulgar language was not used. R spit [sic] at his opponents many times. He used a strong arm whenever he could free himself, and his blows were beyond the ordinary strength of the boy [Bishop 1949, 258].

The exorcism continued on and off for days. At times R screamed "in [a] diabolical, high-pitched voice"; he swung his fists, once breaking Halloran's nose; he sat up and sang (for example, "The Blue Danube," "Old Rugged Cross," and "Swanee"); he cried; he spat; he cursed his father; he mimed masturbation; he bit his caretakers. On March 18, a crisis seemed to occur: as if attempting to vomit, R said, "He's going, he's going . . . " and "There he goes." He went limp and seemed to return to normal. He said he had had a vision of a figure in a black robe and cowl walking away in a black cloud (Bishop 1949, 257–62).

However, after the priests left R claimed to have odd feelings in his stomach and cried out, "He's coming back! He's coming back!" Soon the tantrums and routine of exorcism continued. R seemed even more violent, hurling vulgarities and fists, and he had spells of Satan-dictated writing and speech, for example: "In 10 days I will give a sign on his chest[;] he will have to have it covered to show my power." R also wrote, "Dead bishop" (Bishop 1949, 262–69). Subsequently, on April 1, between disturbances, the youth was baptized in the rectory.

During all this time the markings—the random scratches and words—continued to appear on R's body. When there was talk of his going to school in St. Louis, the boy grimaced and opened his shirt to reveal the scratched words, "No school" (Allen 2000, 46), a seemingly childish concern for truly diabolic forces. (The diary mentions only that "No" appeared on the boy's wrists.)

Reportedly, on one occasion R was observed using one of his fingernails (which were quite long) to scratch the words "HELL" and "CHRIST" on his chest. It is unclear whether he realized he was being observed at the time. Earlier, the priests reportedly "saw a new scratch slowly moving down his leg" (Allen 2000, 180). This sounds mysterious until we consider that the boy could have made a quick scratch just before the priests looked—which they did because he suddenly "yelped"—and what they observed was merely the aftereffect of the scratch, the skin's developing response to a superficial injury. (I have produced just such an effect on myself experimentally, observed by *Skeptical Inquirer*'s Ben Radford.)

On April 4, the family decided to return to their Maryland home, because of the father's need to work and also to relieve the strain on the Missouri relatives. After five days, though, R was sent back to St. Louis and admitted to a hospital run by an order of monks. He was put in a security room that had bars on its single window and straps on the bed. During the day, the teenager studied the catechism and was taken on outings, but at night the "possession" continued. There were failed attempts to give R Holy Communion, "the devil" at one point saying (according to the diary) that he would not permit it (Bishop 1949, 282).

On April 18, R again announced "He's gone!" This time, he said, he had a vision of "a very beautiful man wearing a white robe and holding a fiery sword." With it the figure (presumably Jesus) drove the devil into a pit. There were no further episodes and Father Bishop (1949, 291) recorded that on August 19, 1951, R and his parents visited the monks who had cared for him. "R, now 16, is a fine young man," he wrote. "His father and mother also became Catholic, having received their first Holy Communion on Christmas Day, 1950."

Aftermath

Was R possessed? Or did superstition mask a troubled youth's problems and invite elaborate role-playing? Interestingly, Archbishop Ritter appointed a Jesuit philosophy professor to investigate the matter. According to a reportedly informed source, the investigator concluded that R "was not the victim of diabolical possession" (Allen 2000, 234). Without wishing to make a categorical judgment, Halloran states that R did not exhibit prodigious strength, showing nothing more than what could be summoned by an agitated teenager. As to speaking in Latin, Halloran thought that was nothing more than the boy's having heard repetitious Latin phrases from the exorcising priest. (On one occasion, "the devil reportedly spoke school kids' 'pig Latin'"!)

Nothing that was reliably reported in the case was beyond the abilities of a teenager to produce. The tantrums, "trances," moved furniture, hurled objects, automatic writing, superficial scratches, and other phenomena were just the kinds of things someone of R's age could have accomplished, just as others have done before and since. Indeed, the elements of "poltergeist phenomena," "spirit communication," and "demonic possession"—taken both separately and, especially, together, as one progressed to the other—suggest nothing so much as role-playing involving trickery. So does the stereotypical storybook portrayal of "the devil" throughout.

Writer Mark Opsasnik (2000) investigated the case, tracing the family's home to Cottage City, Maryland (not Mount Rainier as once thought), and talked to R's neighbors and childhood friends. The boy had been a very clever trickster, who had pulled pranks to frighten his mother and to fool children in the neighborhood. "There was no possession," Opsasnik told the *Washington Post*. "The kid was just a prankster" (Saulny 2000).

Of course, the fact that the boy wanted to engage in such extreme antics over a period of three months does suggest that he was emotionally disturbed. Teenagers typically have problems, and R seemed to have trouble adjusting—to school, to his burgeoning sexual awareness, and to other concerns. To an extent, of course, he was challenging authority as part of his self-development, and he was no doubt enjoying the attention. But there is simply no credible evidence to suggest that the boy was possessed by demons or evil spirits.

A Catholic scholar, the Rev. Richard McBrien, who formerly chaired Notre Dame's theology department, states that he is "exceedingly skeptical" of all alleged possession cases. He told the *Philadelphia Daily News* (which also interviewed me for a critical look at the subject), "Whenever I see reports of exorcisms, I never believe them." He has concluded that "in olden times, long before there was a discipline known as psychiatry and long before medical advances . . . what caused possession was really forms of mental or physical illness" (Adamson 2000). Elsewhere, McBrien (1991) has said that the practice of exorcism—and by inference a belief in demon possession—"holds the faith up to ridicule." Let us hope that the enlightened view, rather than the occult one, prevails.

REFERENCES

Adamson, April. 2000. Ancient rite generates modern-day skepticism. *Philadelphia Daily News*, 3 October.

Allen, Thomas B. 2000. *Possessed: The True Story of an Exorcism*. Lincoln, Neb.: iUniverse.com.

Baker, Robert A. 1992. *Hidden Memories: Voices and Visions from Within*. Buffalo, N.Y.: Prometheus Books.

Beyerstein, Barry L. 1988. Neuropathology and the legacy of spiritual possession. *Skeptical Inquirer* 12, no. 3 (Spring): 248–62.

Bishop, Raymond J. 1949. Typescript diary of an exorcism. 25 April (reprinted in Allen 2000, 245–91).

Christopher, Milbourne. 1970. *ESP, Seers & Psychics*. New York: Thomas Y. Crowell.

"The Exorcism." 1991. *20/20*, ABC network broadcast, 5 April.

McBrien, Richard. 1991. Interview on ABC's *Nightline*, 5 April.

Nickell, Joe. 1995. *Entities: Angels, Spirits, Demons, and Other Alien Beings*. Amherst, N.Y.: Prometheus Books, 79–82, 119–20.

———. 1998. *Looking for a Miracle*. Amherst, N.Y.: Prometheus Books.

———. 1999. The Davenport Brothers. *Skeptical Inquirer* 23, no. 4 (July/August): 14–17.

Opsasnik, Mark. 2000. The haunted boy. *Strange Magazine* 20 (serialized on www.strangemag.com/exorcistpage1.html).

Saulny, Susan. 1999. Historian exorcises Mount Rainier's past. *Washington Post*, 24 March.

Vatican updates its rules on exorcism of demons. 1999. *Arizona Daily Star*, 27 January.

4

The "Goatsucker" Attack

Mimicking the "cattle mutilation" hype of yesteryear, during the mid-1990s reports of a bloodthirsty beast—*El Chupacabras*, or "the goat-sucker"—spread from Puerto Rico to Mexico and, still later, to Florida. According to the Cox News Service (April 1996), "The creature supposedly is part space alien, part vampire and part reptile, with long sharp claws, bulging eyes and a Dracula-like taste for sucking blood from neck bites." In Puerto Rico, where the myth originated, "the creature has spawned something near hysteria."

It reportedly attacked turkeys, goats, rabbits, dogs, cats, cows, and horses, sucking the blood from them. However, as Reuters reported, the Puerto Rico Agriculture Department dispatched a veterinarian to investigate. Officials then announced that all the animals had died under normal circumstances and that, contrary to claims, not one had been bled dry (Nickell 1996).

FIGURE 4-1 **Chupacabra action figure**

When the scare spread to Mexico in April of 1996, a scientific team staked out farmyards where the goatsucker had reportedly struck. Wild dogs were caught each time. A police official remarked, "I don't know about the rest of Mexico or the rest of the world, but here the goatsuckers are just dogs." He added: "There is just this huge psychosis. You see it everywhere, even though everywhere we go we prove that there aren't any extraterrestrials or vampires" (Nickell 1996).

As media queries flooded into *Skeptical Inquirer*, I monitored the reports and developments and contacted our colleagues in Mexico City, Patricia and Mario Mendez-Acosta. They interviewed several veterinary pathologists who had conducted numerous necropsies on alleged victims of the goatsucker. Again, in every instance blood was still present in the dead animal.

Some news sources reported that a nurse who lived in a village near Mexico City had been attacked by the goatsucker. Actually, she simply fell and broke her arm, but her cries for help were misinterpreted by her grandmother. Neighbors rushing to her aid saw a black winged form; in reality, it was a flock of swallows, but thus the rumor was born. In another Mexican incident, a man who claimed he had been attacked by the goatsucker later confessed that it was a cover story for his having participated in a brawl (*Los Angeles Times*, May 19, 1996).

Largely through Miami's Latino radio and television stations, the collective delusion has now spread to the Sunshine State. Prompted by local authorities and surrounded by members of the news media, a University of Miami veterinary professor, Alan Herron, cut open a dead goat to demonstrate that it had merely been bitten, not drained of its blood. Citing the bite wounds that were "suggestive of predation," Prof. Herron concluded, "[a] pack of wild dogs did it."

"Of course," reported the Cox News Service, "that did little to calm the chupacabras frenzy" (Nickell 1996).

References

Los Angeles Times. 1996. 19 May.

Nickell, Joe. 1996. Goatsucker hysteria. *Skeptical Inquirer* 20, no. 5 (September/October): 12.

5

Undercover Among the Spirits

Investigating Camp Chesterfield

Camp Chesterfield is a notorious spiritualist enclave located in Chesterfield, Indiana. Dubbed "the Coney Island of spiritualism," it has been the target of many exposés, most notably a book by a confessed fraudulent medium published in 1976 (Keene 1976). A quarter-century later, I decided to see if the old deceptions were still being practiced at the camp. Naturally, my visit was both unannounced and undercover.

The Background

Modern spiritualism began in 1848 with the schoolgirl pranks of Maggie and Katie Fox at Hydesville, New York. Although four decades later the sisters confessed that their "spirit" rappings had been bogus, in the meantime the craze for allegedly communicating with the dead had spread across America, Europe, and beyond. At séances held in darkened rooms and theaters, "mediums" (those who supposedly contacted spirits for others) produced such phenomena as slate writing, table tipping, and "materializations" of spirit entities.

As the number of believers and adherents grew, spiritualist camp meetings became common, and some groups established permanent spiritualist centers. Among these were the Cassadaga Lake Free Association in western New York, founded in 1879 (LaJudice and Vogt 1984). Now known as Lily Dale Assembly, it is the oldest and largest such center in existence. Another, the outgrowth of annual meetings that began in 1886, acquired property at Camp Chesterfield; this center opened with a group of tent dwellings and a large tent "auditorium" in 1891.

In time, many mediums were caught cheating. For example, in the 1860s, Boston "spirit" photographer William H. Mumler was exposed when some of his ethereal images were recognized as living Bostonians. In 1876, another Boston medium was undone when a reporter discovered her confederate, who had played the role of her materialized "spirit guide," hiding in a recess (Nickell 1995, 17–38). The great magician Harry Houdini (1874–1926) spent much of the latter part of his career investigating and exposing mediumistic trickery.

At the spiritualist centers there were many scandals. For instance, one at Lily Dale in 1896 involved a "materializing medium" named Hugh Moore who was caught cheating and arrested. He subsequently jumped bail and, according to a contemporary account, "left his confederates, who helped 'play spirits,' unpaid" (Hyde 1896). Today, though, Lily Dale eschews such trickery and prohibits alleged physical phenomena in favor of pure "mental mediumship."

Perhaps none of the spiritualist centers developed such an unsavory reputation as Camp Chesterfield. Even today, spiritualist friends of mine roll their eyes accusatorily whenever Chesterfield's name is mentioned, and they are quick to point out that the camp is not chartered by the National Spiritualist Association of Churches. The introduction to an official history of Chesterfield (*Chesterfield Lives* 1986, 6), admits that it is surprising the camp has survived, given its troubled past:

> In fact, in its 100 years of recorded history, Camp Chesterfield has been "killed off" more than once! There have been cries of "fraud" and "fake" (and these were some of the nicer things we have been called!) and of course, the "exposés" came along with the regularity of a well-planned schedule. Oh yes! We have been damned and downed—but the fact remains that we must have been doing something right because: CHESTERFIELD LIVES!!

Be that as it may, the part about the exposés is certainly true. A major exposé came in 1960 when two researchers—both sympathetic to spiritualism—arranged to film the supposed materialization of spirits. This was to occur under the mediumship of Edith Stillwell, who was noted for her multiple-figure spirit manifestations, and the séance was to be documented using see-in-the-dark technology. While the camera ran, luminous spectral figures took form and vanished near the medium's cabinet, but when the infrared film was processed the researchers saw that the ghosts were actually fully human confederates dressed in luminous gauze—some were even recognizable as Chesterfield residents. They had not materialized and dematerialized, but rather came and went through a secret door that led to an adjacent apartment (Keene 1976, 40; Christopher 1970, 174).

One of the researchers, Tom O'Neill, himself a devout spiritualist, was devastated by this evidence and railed against "the frauds, fakes and fantasies of the Chesterfield Spiritualist camp!" He added: "The motion picture results of those proceedings will go down in history as the greatest recordings of fraud ever in the movement of Spiritualism. . . . The whole sordid mess is one of the bitterest pills I ever had to swallow" (Keene 1976, 40; Christopher 1970, 175). The other researcher, Andrija Puharich, dubbed Chesterfield "a psychic circus without equal!" (Keene 1976, 41).

An even more devastating exposé came in 1976 with the book *The*

Psychic Mafia, written by former Chesterfield medium M. Lamar Keene. Saying that money was "the name of the game" at Chesterfield, Keene detailed the many tricks used by mediums there, which he dubbed "the Coney Island of Spiritualism." He told how "apports" (said to be materialized gifts from spirits) were purchased and hidden in readiness for a séance; how chiffon became "ectoplasm" (a purported mediumistic substance); how sitters' questions written on slips of paper called *billets* were secretly read and then answered; how trumpets were made to float in the air with discarnate voices speaking through them; and how other tricks were accomplished to bilk credulous sitters (Keene 1976, 95–114).

Keene also told how the billets were shrewdly retained from the various public clairvoyant message services held at Chesterfield. Kept in voluminous files beneath the cathedral, the billets—along with a medium's own private files and those shared by fellow scam artists—provided excellent resources for future readings.

There were still other exposés of Camp Chesterfield. In 1985 a medium from Chesterfield visited Lexington, Kentucky, where he conducted dark-room materialization séances. He featured the production of apports, the floating-trumpet-with-spirit-voices feat, and something called "spirit precipitations on silk." To produce the latter, the sitters' "spirit guides" supposedly took ink from an open bottle and created their own small self-portraits on swatches of cloth the sitters held in their laps.

I investigated when one sitter complained, suspecting fraud. Laboratory analyses by forensic analyst John F. Fischer revealed the presence of solvent stains (shown under argon laser light). Keene gives a recipe for such productions (1976, 110–11), utilizing a solvent to transfer pictures from newspapers or magazines to the materials. This recipe enabled me to create similar "precipitations" (Nickell with Fischer 1988). The prepared swatches had obviously been switched for the blank ones originally shown. With this evidence, as well as affidavits from a few séance victims, I obtained police warrants against the medium on charges of "theft by deception." (These were only misdemeanor charges because—although the medium grossed $800 in an evening—he bilked each victim of only $40. Therefore, we could not extradite him from Indiana, and eventually had to be content with having put an end to his Kentucky séance tricks.)

Undercover

I had long wanted to visit Camp Chesterfield, and in the summer of 2001, following a trip to Kentucky to see my elderly mother and other family members, I decided to head north to Indiana to check out the notorious site (Figures 5-1 and 5-2).

Now, skeptics have never been welcome at Chesterfield. The late Mable Riffle, a medium who ran the camp from 1909 until her death in 1961 (*Chesterfield Lives* 1986) dealt with them summarily. Keene re-

FIGURE 5-1. **Sign at entrance to Camp Chesterfield, dubbed "the Coney Island of Spiritualism."**

FIGURE 5-2. **This scenic view of Camp Chesterfield belies its sordid history of trickery, manipulation, and fraud.**

ported that when Riffle heard one couple using the "f-word"—*fraud*—she snarled, "We do not have that kind of talk here. Now you get your goddam ass off these hallowed grounds and don't ever come back!" (Keene 1976, 48).

Another skeptic, a reporter named Rosie who had written a series of exposés and therefore been banned from the grounds, had the nerve to return to Chesterfield. Wearing a "fright wig," she got into one of Riffle's séances; when the "spirits" began talking through the trumpet, the reporter began to demean them. According to Keene (1976, 48–49), Riffle recognized Rosie's voice immediately and went for her. "Grabbing the reporter by the back of the neck, she ushered her up a steep flight of stairs, kicking her in the rump on each step and cursing her with every profanity imaginable." Nevertheless, the intrepid reporter returned one summer in yet another disguise. Unfortunately she was recognized again, but as Riffle grabbed her, cursing and dragging her toward the gate, some of the camp's financial benefactors arrived. "In a flash," says Keene, "Mable changed her tone. 'Goodbye, Rosie dear,' she said, smiling sweetly, 'we'll be seeing you again some time.'"

With these lessons in mind, I naturally did not want to be recognized at Camp Chesterfield—not out of fear for my personal safety, but so as to be able to observe unimpeded for as long as possible. In my younger years, I was a private investigator with an international detective agency. Then I simply used my own name and appearance (unless I was shadowing a particular individual extensively, in which case I sometimes made some minor adjustments such as using a reversible jacket, donning glasses, etc.). For undercover jobs, I merely wore the attire appropriate for a forklift driver, steelworker, tavern waiter, or other "role" (Nickell 2001).

The same is true of my several previous undercover visits to paranormal sites and gatherings (including a private spiritualistic circle, which featured table-tipping and other séances, that I infiltrated in 2000 [Nickell 2000]). Because I am often the token skeptic on television talk shows and documentaries on the paranormal, I have naturally feared I might be recognized, but I rarely made any effort to disguise myself and usually had no problem. (I was recognized at one "miracle" site in Kentucky, but the television crew accompanying me was able to shush the young man who had announced my presence in a loud voice. At the previously mentioned spiritualist circle, I went unrecognized until I appeared on the television program *48 Hours*, but by then the group had disbanded. Interestingly, word was sent that I would be welcome if the circle was reactivated; supposedly I had "good vibes"!)

However, for my stint at Camp Chesterfield, I felt special measures were called for, so I decided to alter my appearance. I shaved off my mustache (for the first time in more than 30 years!), and replaced my coat-and-tie look with a tee shirt, suspenders, straw hat, and cane. I also adopted a pseudonym, "James Collins," after the name of one of Houdini's assistants. From July 19 to 23, "Jim," who seemed bereft at what he said was the recent death of his mother, limped up and down the grounds and spent nights at one of the camp's two hotels (devoid of such amenities as television and air conditioning). The results were eye-opening, involving a panoply of discredited spiritualist practices that seemed to have changed very little from when they were revealed in *The Psychic Mafia.*

Billet Reading

I witnessed three versions of the old billet scam: one done across the table from me during a private reading, and two performed for church audiences. One of the latter was accomplished with the medium blindfolded.

The first situation, in which the medium works one-on-one with the client, involves getting a peek at the folded slip while the person is distracted. (Magicians call this *misdirection.*) For instance, while the medium directs the sitter's attention—in my case, by pointing to some numerological scribblings—she can surreptitiously open the billet in her lap with a flick of the thumb of her other hand and quickly glimpse the contents. As expected, the alleged clairvoyant knew exactly what was penned on my slip: the names of four persons who had "passed into spirit" and two questions. She did *not* know that the people were fictitious.

Her "cold-reading" technique, although rapid-fire, was unimpressive. She did tell me of her feeling that my previous life may have been as a rabbi who perished in the Holocaust, but that could merely have been a clever invention stemming from the date of my birth—during World War II—which she asked for numerological purposes.

One aspect of the reading, which was held in the séance room of the medium's cottage, was particularly amusing. At times she would turn to her right, as if acknowledging the presence of an invisible entity, and say "Yes, I will." This was a seeming acknowledgment of some message she had supposedly received from a spirit, which she was to impart to me. I paid the medium $30 cash and considered it a bargain—though not in the way the spiritualist would no doubt have hoped!

At both of the billet readings I attended that were conducted for audiences (one in a chapel, the other in the cathedral), a volunteer stood inside the doorway and handed each attendee a slip of paper. Printed instructions at the top directed us to "Please address your billet to one or more loved ones in spirit, giving first and last names. Ask one or more questions and sign your full name."

On the first occasion, I made a point of seeming uncertain about how to fold the paper. I was instructed that it was to be simply doubled over and creased; if it were done otherwise, I was told sternly, the medium would not read it. I did not ask why, as I was trying to appear as credulous as possible, but in fact I knew that there were two reasons. First, of course, the billets needed to be easy to open with a flick of the thumb; second, it was essential that they all look alike. The reason for the latter condition lay in the method employed: After the slips were gathered in a collection plate and dumped atop the lectern (where they could not be seen from our vantage point), the medium would pick one up and hold it to his or her forehead while divining its contents. The trick involves secretly glancing down at an open billet. A sitter who had closed his slip in a distinctive way (such as by pleating it or folding it into a triangle) might notice that the billet being shown was not the one apparently being viewed clairvoyantly.

The insistence on a particular method of paper folding is a red-flag indication of trickery. That was confirmed for me at one session at which I wrote the names of nonexistent loved ones and signed with my pseudonym. From near the back of the chapel, I acknowledged the medium's announcement that he was "getting the Collins family." After revealing the bogus names I had written, he gave me an endearing message, from my supposedly departed mother, that answered a question I had addressed to her on the billet. However, my mother was actually among the living and, of course, is not named Collins.

The other public billet reading I attended was part of a gala service held in the cathedral. The medium placed adhesive strips over her eyes and followed that with a scarf tied in blindfold fashion. This is obviously supposed to prove that the previously described method of billet reading was not being employed, but according to Keene (1976, 45), who had performed the same feat, "The secret here was the old mentalist standby: the peek down the side of the nose." He adds: "No matter how securely the eyes are blindfolded, it's always possible to get enough of a gap to read material held close to the body." Unfortunately, at this reading my billet was not among those chosen, so I received no special communications from the nonexistent persons whose names I had penned.

Even more impressive billet-reading tricks are described by Keene (1976, 97–98), including the feat of reading billets sealed in envelopes, or reading one or two dramatically as the tray or basket is being carried to the medium. However, I saw none of these more sensational (but less convenient to perform) methods.

Spirit Writing

Another feat practiced by at least three mediums at Chesterfield is called *spirit card writing*. This trick is a descendant of the old slate effects that were common during the heyday of spiritualism, whereby (in a typical effect) alleged otherworldly writing mysteriously appeared on the inner surfaces of a pair of slates that were bound together (Nickell 2000). In the modern form (which boasts several variants), blank cards are placed in a basket along with an assortment of pens, colored pencils, and other marking implements. After a suitable invocation, each of the cards is seen typically to bear a sitter's name, surrounded by the names of his "spirit guides" or other entities and possibly a drawing or other artwork. The sitter keeps the card as a tangible "proof" of spirit power.

At Chesterfield, I attempted to sign up for a private card-writing séance one evening at the home of a prominent medium (who also advertises other feats, including "pictures on silk"). When that session proved to be filled, I decided to try to "crash" the event and soon hit on a subterfuge. I placed the autograph of "Jim Collins" on the sign-in sheet for the *following* week, then showed up at the appointed time for the current séance a few hours later. I milled about with the prospective sitters, and then we were all ushered into the séance room in the medium's bungalow.

So far so good. Unfortunately, when the medium read off the signees' names and found that I was unaccounted for, I had some explaining to do. I insisted that I had signed the sheet and let him discover the "error" I had made. Then, suitably repentant and deeply disappointed, I implored him to allow me to stay, noting that there was more than one extra seat. Of course, if the affair were bogus, and the cards prepared in advance, I could not be permitted to participate. Not surprisingly, I was not, though I was given the lame excuse (by another medium, a young woman, who was sitting in on the session) that the medium needed to prepare for the séance by "meditating" on each sitter's name. (I wondered which of the two types of mediums she was: one of the "shut-eyes," simple believers who fancy that they receive psychic impressions, or one of the "open mediums," who acknowledge their deceptions within the secret fraternity [Keene 1976, 23].) Even without my admission fee, I estimate the medium grossed approximately $450.

The next day I sought out one of the sitters, who consoled me over my not having been accepted for the séance. She showed me her card, which bore a scattering of names like "Gray Wolf" in various colors of felt-pen handprinting—all appearing to me, on brief inspection, to have been done by one person. The other side of the card bore a picture (somewhat resembling a Japanese art print) that she thought had also been produced by spirits, although I do not know exactly what was claimed by the medium. I did examine the picture with the small lens on my Swiss Army knife, which revealed the telltale pattern of dots from the halftone printing process. The woman seemed momentarily discomfited when I showed her this, and indeed acknowledged that the whole thing seemed hard to believe, but she stated that she simply *chose* to believe. I nodded understandingly; I was not there to argue with her.

Given the evidence, I have no doubt that blank cards were switched for prepared ones, a deception that could be easily accomplished. In fact, Keene (1976, 109–10) describes several variations of the trick, which he himself performed at Camp Chesterfield.

"Direct Voice"

My most memorable—and unbelievable—experience at Camp Chesterfield involved a spirit materialization séance I attended at a medium's cottage on a Sunday morning. Such offerings are not scheduled in the camp's guidebook, but rather are advertised via a sign-in book, and perhaps an accompanying poster, on the medium's porch. As my previous experience showed, it behooved one to keep abreast of the various offerings around the village. Thus I was out early in the morning, hobbling with my cane up and down the narrow lanes. Soon a small poster caught my eye: "Healing Séance with Apports." It being just after 6:00 A.M., and the streets silent, I quietly stepped onto the porch and signed up for the 10:00 A.M. session.

At the appointed time, seven of us had gathered, and the silver-haired medium ushered us into the séance room. She promptly secured the room against light leakage, placing a rolled-up throw rug at the bottom of the outer door and another rolled cloth to seal the top, and closing a curtain across an interior door. She collected $25 from each attendee and then, after a brief prayer, launched into the healing service. This consisted of a "pep talk" (as she termed it) followed by a brief session with each participant in which she clasped the person's hands and supposedly imparted healing energies.

It eventually came time for the séance. A pair of tin spirit trumpets standing on the floor by the medium's desk indicated that we might experience "direct voice," by which spirits supposedly speak, often using such trumpets to amplify their vocalizations. The medium began by turning off the lamps and informing us that "dark is light." Soon, in the utter blackness, the voices came, seeming to be speaking in turn through one of the trumpets. Keene (1976, 104–8) details various means of producing "levitating" trumpets, complete with luminescent bands around them "so that the sitters could see them whirling around the room, hovering in space, or sometimes swinging back and forth in rhythm with a hymn." This time, however, we were left to our imaginations. Mine suggested to me that the medium was not even bothering to use the large trumpet, which might prove tiresome, but may have been utilizing a small tin megaphone—another trick described by Keene.

Some mediums were better at pretending direct voice than others. Sometimes, according to one critic, "[a]ll the spirit voices sounded exactly like the medium" (Keene 1976, 122). Such was the case at my séance. The first voice sounded just like the medium using exaggerated enunciation to simulate an "Ascended Master" (who urged the rejection of negativity); another sounded just like the medium adopting the craggy voice of "Black Elk" (with a message about having respect for the Earth); still another sounded just like the medium using a perky little-girl voice to conjure up "Miss Poppy" (supposedly one of the medium's "joy guides").

At the end of the séance, after the lights were turned back on, one of the trumpets was lying on its side on the floor, as if dropped there by the spirits—or, as I thought, simply tipped over by the medium. Finally, we were invited up to get our "apports."

Apports

Supposedly materialized or teleported gifts from the spirits, *apports* appear at some séances under varying conditions—sometimes tumbling out of a spirit trumpet, for example. Keene (1976, 108) says those at Camp Chesterfield were typically "worthless trinkets" such as brooches or rings often "bought cheap in bulk." One medium specialized in "spirit jewels" (colored glass), whereas another apported arrowheads; special customers might receive something "more impressive." Camp Chesterfield instructed its apport mediums to "*please ask your guides* to bring articles of equal worth to each sitter and not to bring only one of such articles as are usually in pairs (earrings or cufflinks, for instance)" ("The Medium's Handbook," quoted in Keene 1976, 63).

At our séance, the apports were specimens of hematite, which (like many other stones) has a long tradition of being considered to have healing and other powers (Kunz 1913, 6, 80–81). The shiny, steel-gray mineral had obviously been tumbled (mechanically polished), as indicated by surface characteristics shown by stereomicroscopic examination, and was indistinguishable from specimens purchased in shops that sell such New Age talismans.[1]

The medium handed each of us one of the seven stones, after picking it up with a tissue and noting with delight our reactions at discovering that they were icy cold! This was a nice touch, I thought, imparting an element of unusualness as if somehow consistent with the stones' having been materialized from the Great Beyond—although they were probably only kept by the medium in her freezer until just before the séance, when they were likely transferred to a thermos jar. We were told that each apport was attuned to that sitter's own energy "vibrations" and that no one else should ever be permitted to touch it. If someone did, we were warned, it would become "only a stone."

I left Camp Chesterfield on the morning of my fifth day there, after first taking photographs around the village. As I reflected on my experiences, things seemed to have changed little from the time Keene wrote about in *The Psychic Mafia*. Indeed, the deceptions practiced during my stay there harkened back to the days of Houdini and beyond—actually, all the way back to 1848, when the Fox sisters launched the spiritualist craze with their schoolgirl tricks.

NOTE

1. I identified the "apport" as hematite (Fe_2O_3) by its steel-gray luster, Mohs'-scale hardness, dark-red streak-plate color, and other properties, including demonstrated iron composition. (This was shown by acid digestion of the streak trace followed by application of potassium ferrocyanide reagent, which yielded the prussian-blue color reaction that indicates the presence of iron.) Stereomicroscopic observation showed surface characteristics comparable to known specimens of tumbled hematite, including pitting and fine striations.

REFERENCES

Chesterfield Lives. 1986.Chesterfield, Indiana: Camp Chesterfield.

Christopher, Milbourne. 1970. *ESP, Seers & Psychics*. New York: Thomas Y. Crowell.

Hyde, Julia. 1896. Letter to a Mrs. Peck, written at Lily Dale, 25 April. [Typescript text in file on the Hyde house, Genesee Country Village, Mumford, N.Y.]

Keene, M. Lamar. [1976] 1997. *The Psychic Mafia*. Amherst, N.Y.: Prometheus Books.

Kunz, George Frederick. [1913] 1971. *The Curious Lore of Precious Stones*. New York: Dover.

LaJudice, Joyce, and Paula M. Vogt. 1984. N.p. [Lily Dale, N.Y.]: n.p. [Lily Dale Assembly].

Nickell, Joe, with John F. Fischer. 1988. *Secrets of the Supernatural*. Buffalo, N.Y.: Prometheus Books.

———. 1995. *Entities*. Amherst, N.Y.: Prometheus Books.

———. 2000. Spirit painting, part II. *Skeptical Briefs* 10, no. 2 (June): 9–11.

———. 2001. Adventures of a paranormal investigator. In *Skeptical Odysseys*, edited by Paul Kurtz, 219–32. Amherst, N.Y.: Prometheus Books.

6

Alien Hybrid?

In a cabinet in a small natural-history museum in Waldenburg, Saxony, is a strange curio (Figure 6-1) that one writer has termed "Germany's greatest mystery" (Hausdorf 2000). It is the fetus of—well, that is the question: What is it? Could it be, as some ufologists insist, an alien hybrid?

Known locally from its strange appearance as the "chicken man," the fetus can be traced to the year 1735, when it was stillborn in the Saxonian village of Taucha. It was to have been the fourth child of Johanna Sophia Schmiedt, who was in the eighth month of her pregnancy. She was 28, and her husband Andreas, "a hunchback," was 10 years her senior. Two years later, in 1737, Leipzig physician Gottlieb Friderici autopsied the preserved fetus and published a report, illustrated with two copperplate engravings and titled "Monstrum Huma-

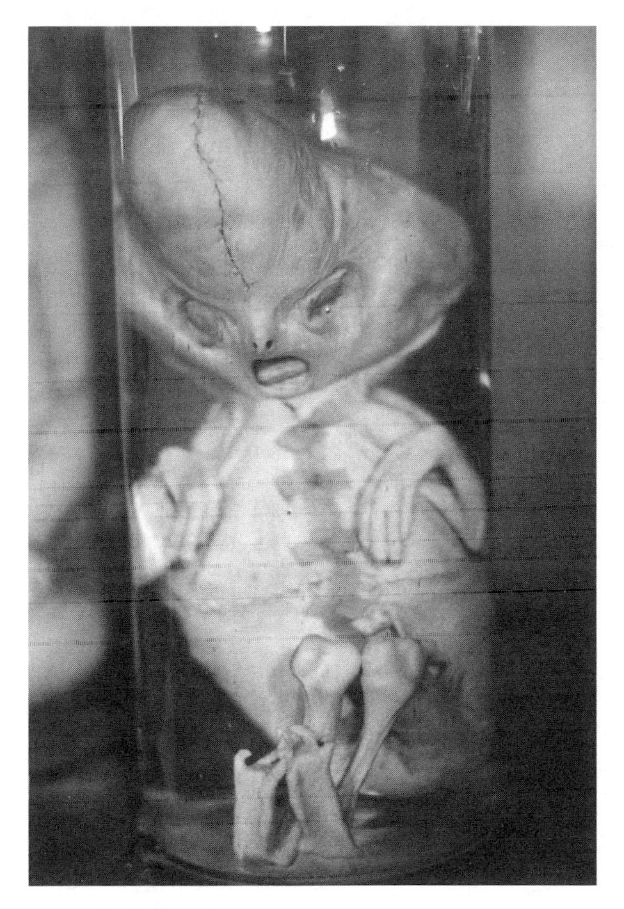

FIGURE 6-1. **Alleged "alien hybrid" in Saxony.**

num Rarissimum" ("Most Rare Human Monster") (Müller 1999; Monstrum 1994; Ausserirdisches 2002).

Dr. Friderici concluded that the term "monster" was appropriate because of the fetus's divergence from normal human anatomy. According to one treatise on monsters, "From the earliest period of the world's history abnormal creatures or monstrosities, both human and animal, have existed from time to time and excited the wonder of mankind."

The births of monsters were explained in superstitious, often supernatural, terms. They have been thought to presage calamities and disasters, or to be evidence of divine wrath. Some believed them to be the result of mating with animals (Thompson 1968, 17–24).

A widespread popular notion was that (like birthmarks and other defects) monsters were caused by something the mother saw or touched during her pregnancy (DeLys 1989, 219–20). In fact, in the case at hand, Mrs. Schmiedt's patient history recorded her own apparent explanation: she previously had had a very frightening encounter with a marten (an animal related to the weasel).

In mentioning this, one writer suggests that the recollected marten was only a "cover memory" for an extraterrestrial encounter, and claims that the deformed fetus was an alien/human hybrid. Supposedly it resembles the small "grays" (Ausserirdisches 2002), but the comparison is poor. For example, although the allegedly humanoid gray is portrayed with a large head, it lacks the bizarre bulbous growth of this fetus. Also, whereas the grays supposedly have a "distinguishing characteristic [of] black, wraparound eyes" (Huyghe 1996, 13–16), the Waldenburg fetus instead possesses "very *round* eye sockets" (Ausserirdisches 2002 [emphasis added]).

One source claims of the fetus that "nothing in either its interior or exterior configuration corresponds to that which is considered to be human" (Monstrum 2002), but that is a ridiculous exaggeration. Not only does it have an essentially human body structure (a head atop a cylindrical trunk, with arms attached to the shoulders and legs extending from the hips), but the interior was also found to contain such organs as a heart, liver, and lungs, according to Dr. Friderici's autopsy results. The anatomy is also characterized by a vertebrate skeletal structure, although the lower arm and leg bones did not each become differentiated into two separate bones; the cranium had, in effect, burst open, and some of the skull bones (e.g., the upper jawbone) were missing.

Fringe paranormalists attribute the abnormalities to the purported hybridization. One Internet source bizarrely claims to have received inside information on the matter from "the Cassiopaeans," allegedly channeled extraterrestrial entities. "They" explained the fetus as "hybridized conception/gestate" and averred: "It was not an experiment. It was the result of Reptoid 'rape'" (Monstrum 2002).

Actually, the Cassiopaeans are out of touch with reality. Visiting the museum on a trip to Germany in 2002 with Martin Mahner, who has a doctorate in zoology, I was able to learn the facts about the "chicken man." Museologist Ulrike Budig was most helpful, unlocking the cabinet's glass doors so we could examine and photograph the specimen from various angles.

We learned that genetic tests had been conducted by experts in Berlin and Heidelberg (Monstrum 1994). They studied the chromosomes using comparative genomic hybridization, an analytical method developed in 1993. This showed that the fetus was female (two X chromosomes being present). More importantly, the scientists discovered that large parts of chromosome 17 were missing, and concluded that a genetic loss of that amount is more than likely the cause of the rare deformation. Because no other case of this particular chromosome-17 anomaly is known, one must assume that the result is lethal, meaning that such embryos normally die at an early stage and are spontaneously miscarried. In this unique instance, however, further development of the embryo took place until the third trimester of the pregnancy (Müller 1999).

Dr. Dietmar Müller (1999), writing in the museum's official guidebook, observes sagely that, as occurred with folk of the eighteenth century, such deformations have also inspired the fantasies of some modern people. In this instance, he notes, "ufologists" have brought "aliens" into play, thus giving the museum specimen an unfortunate notoriety. However, he concludes that the scientific studies clearly prove that this specimen has only human DNA.

REFERENCES

Ausserirdisches leben—Monstrum humanum rarissimum. Last accessed 20 August 2002. http://acolina.grenzwissen.de/content/seti/huhnm.htm.

DeLys, Claudia. 1989. *What's So Lucky About a Four-leaf Clover?* New York: Bell Publishing.

Hausdorf, Hartwig. 2000. Monstrum humanum rarissimum. *Fate*, 1 February, 28–32.

Huyghe, Patrick. 1996. *The Field Guide to Extraterrestrials*. New York: Avon Books.

Monstrum humanum. 1994. *Der Spiegel*, 27 (4 July): 165.

Monstrum humanum rarissimum. Last accessed 14 November 2002. http://cassiopaea.xmystic.com/en/cass/monster.htm.

Müller, Dietmar. 1999. "Monstrum humanum"—Die anatomische Sammlung. In Naturalienkabinett Waldenburg, edited by Ulrike Budig et al., 91–98.Waldenburg, Saxony, Germany: Heimatmuseum und Naturalienkabinett.

Thompson, C. J. S. 1968. *Giants, Dwarfs and Other Oddities*. New York: Citadel Press.

7

Image of Guadalupe

Myth-perception

A radio announcer asked his listeners to brace themselves for a report that would "shock all of Mexico." When it came, it did not announce an assassination, as some had assumed, but for many it was even worse. It concerned the most popular shrine in the Roman Catholic world next to the Vatican. As the San Antonio *Express-News* reported in its five-column headline: "Faithful aghast as abbot paints Virgin story as myth" ("Faithful" 1996).

The reference is to the tale that in 1531 (some 10 years after Cortez's defeat of the Aztec Empire), the Virgin Mary appeared to an Indian peasant named Juan Diego and gave him a "sign": her full-length image miraculously imprinted on the inside of his cloak. The reputed miracle was instrumental in effecting the mass conversion of the Aztecs to Catholicism (Smith 1983).

Now, though, Monsignor Guillermo Schulemburg, abbot of the

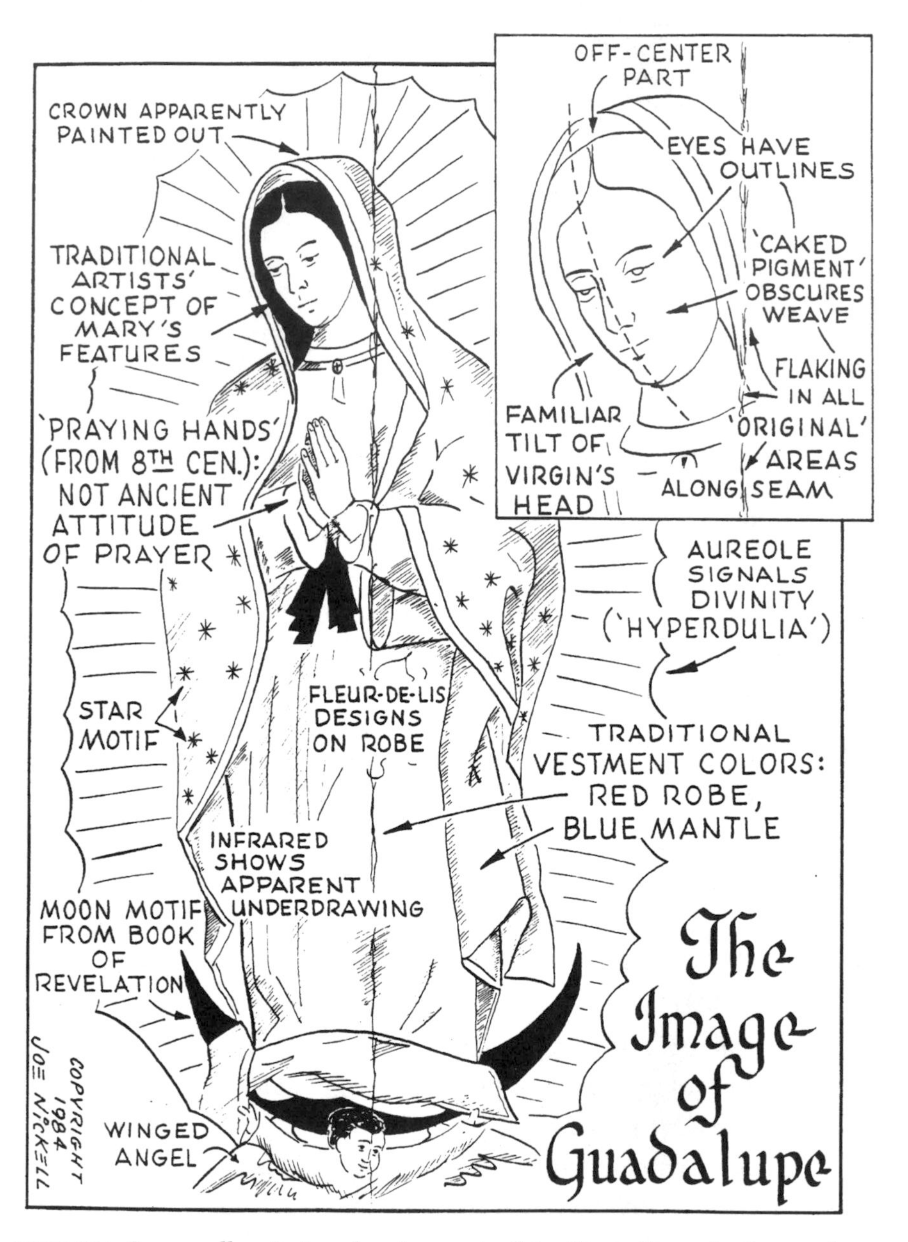

FIGURE 7-1. **Supposedly a "miraculous" portrait of the Virgin Mary, the Image of Guadalupe actually exhibits artistic motifs and evidence of painting.**

Basilica of Guadalupe where the Image is enshrined, was admitting that the whole story was a myth. The Italian magazine *30 Giorni* ("30 Days") quoted Schulemburg as saying that Juan Diego was fictitious, "a symbol, not a reality" (cited in "Faithful" 1996). He also stated, according to the magazine, that Juan Diego's 1990 beatification by Pope John Paul II (a step preparatory to sainthood) was simply "a recognition of a cult. It is not a recognition of the physical, real existence of the person" (cited in "Faithful" 1996).

In an extensive folkloristic and iconographic investigation of the Image, forensic analyst John F. Fischer and I learned that the Guadalupan story was quite similar to an earlier Spanish legend, and that the portrait of the Virgin in the Image was typical of Spanish art of the period. Although obvious evidence of paint throughout the Image areas has prompted miraculists to claim that the paint was added later, infrared photographs indicate otherwise, revealing apparent preliminary sketch lines and other evidence that the picture was produced in the usual manner for a painting. (See Figure 7-1.) Moreover, during a formal investigation of the cloth in 1556, one priest testified that the Image had been "painted yesteryear by an Indian," and another that it was "a painting that the Indian painter Marcos had done." An Aztec painter, Marcos Cipac, was active in Mexico at the time the Image of Guadalupe appeared (Smith 1983, Nickell and Fischer 1985).

As the faithful reacted bitterly to Schulemburg's reported statements, the office of the beleaguered, 80-year-old abbot issued a statement saying the comments attributed to him were "absolutely false." However, my colleague, *Skeptical Inquirer* magazine's Mexico City correspondent, Patricia Lopez Zaragoza, provided this informed assessment:

> Schulemburg has repeatedly said that the apparitions are a myth. He has pointed this out since some years ago. After this scandal Schulemburg insisted a couple of times on his version of the story. However, after political pressure grew

> against him he somewhat retreated to a safer position, claiming the value of the apparitions as a mere symbol of Mexican Catholicism.

"It must be said," she added, "that he hasn't really retracted his previous antiapparitionist position."

In early September 1996, the Roman Catholic Church announced Abbot Schulemburg's resignation as head of the basilica. According to an Associated Press dispatch, the church "gave no reason for his departure."

In 2002, the results of a secret 1982 scientific study of the Image were reported by the Spanish-language magazine *Proceso* (in its May 12 and 19 issues). Art restoration expert José Sol Rosales examined the cloth with a stereomicroscope and determined that it had not originated supernaturally, but was instead the work of an artist who used the materials and methods of the sixteenth century.

According to Rosales, the canvas appeared to be a mixture of linen and hemp or cactus fiber. It was prepared with a brush coat of white primer (calcium sulfate), and the painting was then rendered in distemper (i.e., paint consisting of pigment, water, and a binding medium). The artist used a "very limited palette," stated the expert, consisting of black (from pine soot), white, blue, green, various earth colors ("tierras"), reds (including carmine), and gold (Vera 2002a; Vera 2002b).

Rosales's report confirms and amplifies what skeptics had determined from early records, infrared photographs, and other evidence (Nickell and Fischer 1985). In addition, new scholarship suggests that, whereas the Image was painted not long after the Spanish conquest (when miraculous powers were almost immediately attributed to it), the pious legend of Mary's appearance to Juan Diego may date from the following century (Vera 2002c; Poole 1996).

Meanwhile, none of this appears to have had any effect on the Vatican, which proceeded to canonize "Juan Diego" as a saint, fictitious or not.

REFERENCES

Faithful aghast as abbot paints Virgin story as myth. 1996. San Antonio, Texas. *Express-News*, 2 June.

Nickell, Joe, and John F. Fischer. 1985. The Image of Guadalupe: A folkloristic and iconographic investigation. *Skeptical Inquirer* 9, no. 3 (Spring 1985): 243–55.

Poole, Stafford. 1996. *Our Lady of Guadalupe*. Tucson: Arizona University Press.

Smith, Jody Brant. 1983. *The Image of Guadalupe*. Garden City, N.Y.: Doubleday.

Vera, Rodrigo. 2002a. Manos humanas pintaron la Guadalupana. *Proceso*. May 12, 27–30.

———. 2002b. El análisis que ocult el Vaticano. *Proceso*, May 19, 28–31.

———. 2002c. Hacer la biografia del supuesto pintor de la Guadalupana. *Proceso*, June 2, 66–68.

8

Human Blowtorch

This case involves a young African-American man with a seemingly remarkable ability. It has interested a string of mystery mongers, from Charles Fort (1932) to Frank Edwards (1961), Vincent Gaddis (1967), and, more recently, Jane Goldman (1995, 132). Edwards called his account "Human Blowtorch"—a "unique case," he said, for which "no one ever proposed any explanation" (1961, 163–64).

A contemporary account was published in 1882 in *Michigan Medical News* and is given here in its entirety:

> A Singular Phenomenon.—Dr. L. C. Woodman, of Paw Paw, Mich., contributes the following interesting though incredible observation: I have a singular phenomenon in the shape of a young man living here, that I have studied with much interest and I am satisfied that his peculiar power demon-

> strates that electricity is the nerve force beyond dispute. His name is Wm. Underwood, aged 27 years and his gift is that of generating fire through the medium of his breath assisted by manipulations with his hands. He will take anybody's handkerchief and hold it to his mouth[,] rub it vigorously with his hands while breathing on it and immediately it bursts into flames and burns until consumed. He will strip and rinse out his mouth thoroughly, wash his hands and submit to the most rigid examination to preclude the possibility of any humbug, and then by his breath blown upon any paper or cloth envelop it in flame. He will, when out gunning and without matches desirous of a fire, lie down after collecting dry leaves and by breathing on them start the fire and then coolly take off his wet stockings and dry them.

The account continues:

> It is impossible to persuade him to do it more than twice in a day and the effort is attendant with the most extreme exhaustion. He will sink into a chair after doing it, and on one occasion after he had set a newspaper on fire as narrated, I placed my hand on his head and discovered his scalp to be violently twitching as if under intense excitement. He will do it anytime, no matter where he is, under any circumstances, and I have repeatedly known of his setting back from the dinner table, taking a swallow of water and by blowing on his napkin at once set it on fire. He is ignorant and says that he first discovered his strange power by inhaling and exhaling on a perfumed handkerchief that suddenly burned while in his hands. It is certainly no humbug, but what is it? Does physiology give a like instance, and if so when?

What are we to make of this? We can scarcely trust Dr. Woodman's assurance that the phenomenon was "certainly no humbug." He seems one of those (like Sir Arthur Conan Doyle) who believes that he is too smart to be fooled; hence, because he saw no trickery, there could have been none. (Houdini knew that the simplest card trick could fool Sir Arthur.) In response to such people, I have a saying: "The person who thinks he can't be fooled has just fooled himself."

Gaddis did concede that legerdemain was possible in the case. "It is true," he wrote, "that a trickster with nimble fingers, a piece of phosphorus and a bit of saliva might possibly duplicate these feats" (1967, 173–74). However, Gaddis thought that in light of alleged "poltergeist fire-makers," one should not be too quick to dismiss Underwood's ability as mere deception. (For a debunking of poltergeists, see Christopher 1970.)

Gaddis may have chanced to read Houdini's book *Miracle-Mongers and Their Methods* (1920, 105–7), which has this to say:

> To set paper on fire by blowing upon it, small pieces of wet phosphorus are taken into the mouth, and a sheet of tissue paper is held about a foot from the lips. While the paper is being blown upon[,] the phosphorus is ejected on it, although this passes unnoticed by the spectators, and as soon as the continued blowing has dried the phosphorus it will ignite the paper.

Of course, if Underwood employed this method, he would have had to secretly store the phosphorus and surreptitiously introduce it into his mouth—achieving this in much the same way that a strip-searched Houdini might have done with a lock pick (Gibson 1930). It is tempting to think that Underwood might have utilized the very water with which he rinsed his mouth as a means of keeping a pellet of phosphorus protectively wet, and then—after the searching and rinsing—

transferring it in a follow-up sip to the tongue. Underwood's "intense excitement" might only have been due to his natural nervousness and fear of being caught at deception. That he would not perform the feat more than twice in one day merely suggests that some preparation was necessary to repeat the effect. Or it may indicate a magician's shrewdness: conjurers know that repeating a trick increases the risk of discovery. "Once is magic," we magicians say; "twice is an education."

Another method involves use of a hollowed-out nut filled with smoldering punk (a material that, when lighted, will smolder and is often used to ignite fireworks). This method is attributed to a fire-breathing wonder-worker from ancient Roman times, a Syrian called Eunus (d. 133 B.C.E.). To excite his fellow slaves to revolt against the Romans, Eunus claimed that he had received supernatural powers from the gods, who foretold that he would someday be king. As proof, Eunus exhaled jets of fire like the legendary dragon (Nickell 1991, 13–14).

In his book on carnival life, titled *Step Right Up!*, fire-eater/sword-swallower Dan Mannix (1951, 149–50) relates a humorous story of his experimentation with this "very old" stunt. To practice, he used to slip the nut into his mouth after breakfast "and carry it all day long like a quid of tobacco," blowing out sparks and smoke every now and then to keep it lighted. For a sideshow act, "you begin to eat cotton or paper and secretly slip the nut in your mouth with the cotton."

One night, when out with a friend, Mannix stopped in a drugstore and somewhat absentmindedly leaned on the counter while, he wrote, "thoughtlessly rolling the nut around in my mouth." Suddenly he burped, ejecting a burst of smoke and sparks, some of which landed on the hands of the clerk who was wrapping his purchase. The firebreather clapped his hand over his mouth, exclaimed, "I beg your pardon," and made a hasty exit with his package. "As we drove off," Mannix recalled, "I happened to glance back at the store. The clerk and his few customers were standing around staring at each other as though suddenly paralyzed."

Mannix's anecdote indicates that the smoldering-nut feat was long known to carnys. Indeed, the very same "deception of breathing out flames, which at present excites," is described by Houdini (1920, 117–18), who details various improvements on the ancient trick.

Other techniques that "Human Blowtorch" William Underwood might have used to set paper, cloth, and dry leaves aflame also come to mind, including a chemical spontaneous-combustion technique (see Nickell 1998), and still other means. Whatever the exact method—and the phosphorus trick might be the most likely—the possibilities of deception far outweigh any "occult powers" hinted at by Charles Fort (1932, 926) or others.

REFERENCES

Christopher, Milbourne. 1970. *ESP, Seers & Psychics*. New York: Thomas Y. Crowell.

Edwards, Frank. 1961. *Strange People*. New York: Signet.

Fort, Charles. [1932] 1974. *Wild Talents*, reprinted in *The Books of Charles Fort, Part 2*. Ann Arbor, Mich.: University Microfilms.

Gaddis, Vincent H. 1967. *Mysterious Fires and Lights*. New York: David McKay.

Gibson, Walter B. 1930. *Houdini's Escapes and Magic*. New York: Blue Ribbon Books.

Goldman, Jane. 1995. *The X-Files Book of the Unexplained*. New York: Simon & Schuster.

Houdini, Harry. [1920] 1980. *Miracle-Mongers and Their Methods*. Reprinted Toronto: Coles Publishing.

Mannix, Dan. 1951. *Step Right Up!* New York: Harper & Brothers.

Nickell, Joe. 1991. *Wonder-workers!* Buffalo, N.Y.: Prometheus Books.

———. 1998. Fiery tales that spontaneously destruct. *Skeptical Inquirer* 22, no. 2 (March/April): 62.

A singular phenomenon. 1882. *Michigan Medical News* (Detroit) 5: 263.

9

Remotely Viewed?

The Charlie Jordan Case

Was fugitive drug smuggler Charlie Jordan nabbed after a CIA "remote viewer" helped pinpoint his location in northern Wyoming? Does this supposedly successful use of psychic phenomena (or "psi") by the intelligence agency's then-secret Stargate program indeed represent "one of their more memorable cases" (*Mysteries* 1998)?

For an episode of the television series *Mysteries* (which subsequently aired in England on November 23, 1998), I was asked by the British Broadcasting Corporation (BBC) to examine and comment on claims made about the case. Not surprisingly, I found much more information than my brief on-air time allowed me to relate, and I have since learned even more about the matter.

Stargate

The story of the Jordan case, which involved the psychic-spy program in 1989, actually has its roots in Cold-War-era paranoia. In 1970, the book *Psychic Discoveries Behind the Iron Curtain* touted the Soviet Union's allegedly "significant breakthroughs in psychic research"—a field, the authors noted, that was "usually ignored by Western science" (Ostrander and Schroeder 1971, xix). In 1972, U.S. analysts from the Defense Intelligence Agency (DIA) issued a report warning that Soviet "psi research" might eventually permit the adversaries to learn the contents of secret documents, divine the movements of troops and ships, discern the location and purpose of installations, and even "mould the thoughts" of American leaders or—presumably through the reputed powers of psychokinesis (mind over matter)—possibly to "cause the instant death of any US official at a distance" or remotely disable "US military equipment of all types, including spacecraft" (Lamothe 1972, 94–95; Schnabel 1995, 10).

Late in 1972, the Central Intelligence Agency (CIA) provided an initial study grant of $50,000 to a California think-tank called Stanford Research Institute. SRI was to determine whether there was any validity to a form of alleged extrasensory perception (ESP) termed "remote viewing." When the program failed to show promise, despite an annual budget that expanded to between $500,000 and $1 million, the CIA abandoned it in the late 1970s. However, the DIA soon took over and revived the program, operating it as a secret project code-named Stargate until it was finally suspended in 1995 and declassified.

Stargate had three components: one attempted to track other countries' psychic warfare projects; another provided six (later only three) "remote viewers" to any government agency desiring to use them; and the third continued the laboratory research initiated at SRI (and subsequently transferred to another think-tank at Palo Alto, California, called Science Applications International Corporation) (Schnabel 1997, 10–11, 380; Hyman 1996).

The term *remote viewing* was coined by Harold Puthoff, who, with his colleague Russell Targ, ran the psi project at SRI. The alleged ability to see distantly using psi is known as *clairvoyance* ("clear seeing"). That term, however, seemed dated and loaded with undesirable connotations, evoking gypsy fortunetellers, spiritualist mediums, and flaky visionaries. Instead, Targ and Puthoff experimented with more modernesque, technological-sounding terms, until Puthoff's "remote viewing" caught on. (Targ suggested that "remote sensing" would be more accurate, because senses other than sight may supposedly be involved as well.) (Guiley 1991, 111–13; Schnabel 1997, 141–56).

Remote viewing (RV) is actually a more specific term than *clairvoyance*, describing a particular type of alleged psychic sensing called "traveling clairvoyance" or, alternately, "telesthesia." As distinguished from some other types of clairvoyance, such as retrocognitive and precognitive perception (alleged past and future sensing, respectively), *remote viewing* describes "seeing remote or hidden objects clairvoyantly with the inner eye, or in alleged out-of-body travel." Said to be "one of the oldest and most common forms of psi," it is also "one of the most difficult to explain" (Guiley 1991, 468–69).

The best of the secret remote-viewing experiments involved a group of three "gifted" viewers, all of whom were involved in each experiment. Typically,

> [t]he remote viewer would be isolated with an experimenter in a secure location. At another location, a sender would look at a target that had been randomly chosen from a pool of targets. The targets were usually pictures taken from the *National Geographic*. During the sending period the viewer would describe and draw whatever impressions came to mind. After the session, the viewer's description and a set of five pictures (one of them being the actual target picture) would be given to a judge. The judge would then decide which picture was closest to the viewer's description. If the actual target was judged closest to the description, this was scored as a "hit" [Hyman 1996].

With this procedure, one hit could be expected by chance in 20 percent of the attempts. A consistently higher score was considered evidence of psychic ability.

After the experiments were suspended in 1995, the CIA commissioned the American Institutes for Research (AIR) to evaluate the 20-year results. To assess the laboratory component of the remote-viewing research, AIR hired psi-believer Jessica Utts, professor of statistics at the University of California at Davis, and skeptic Ray Hyman, professor of psychology at the University of Oregon and a CSICOP Fellow. The two worked independently and produced separate reports.

Both Utts and Hyman agreed that a group of the 10 best experiments did produce "hit" rates that were consistently above chance. However, they also agreed that the studies were flawed in that they involved a single judge, who was also the main investigator; it remained to be demonstrated that significant scores would still be obtained when independent judges were employed.

Nevertheless, Utts concluded that, taken with other parapsychological experiments, the results were indeed evidence of psychic functioning. Hyman (1996), in contrast, observed that "[t]he history of parapsychology is replete with 'successful' experiments that subsequently could not be replicated." Pointing out that remote viewing and other alleged forms of ESP were defined *negatively*—that is, as an effect remaining after other normal explanations had supposedly been eliminated—Hyman noted that a mere glitch in the experimental data could thus be counted as evidence for psychic phenomena. "What is needed, of course," he said, "is a positive theory of psychic functioning that enables us to tell *when psi is present* and *when it is absent*." He added, "As far as I can tell, every other discipline that claims to be a science deals with phenomena whose presence or absence can clearly be decided."

Other evaluators—two psychologists from AIR—assessed the potential intelligence-gathering usefulness of remote viewing. They concluded that the alleged psychic technique was of dubious value and lacked the concreteness and reliability necessary for it to be used as a basis for making decisions or taking action. The final report found "reason to suspect" that in "some well publicised cases of dramatic hits," the remote viewers might have had "substantially more background information" than was at first apparent (Mumford et al. 1995).

Seeking a Fugitive

Such criticisms are clearly applicable to the Charlie Jordan case. Charles Frank Jordan had been an agent for the U.S. Customs section of the Drug Enforcement Agency in South Florida. Once a trusted employee, who had helped fight drug smuggling in the southern coastal areas of the state and in the Florida Keys, Jordan became a "Customs rotten apple." According to Bill Green, a retired U.S. Customs official, Jordan was found to be "taking bribes to let other people bring drugs in" (*Mysteries* 1998). When he learned that he was under suspicion, he fled, evading an intensive search for two years, during which time he was even featured on the television series *America's Most Wanted* (Graff 2000).

In the spring of 1989, the Customs Service sought the help of the DIA's remote-viewing unit. The psychics "saw" the fugitive in a variety of locales, including south Florida, the Caribbean, and Central America. One, however, supposedly "narrowed down his location" to an area of northern Wyoming. Although the alleged information was never acted on, Jordan was captured some weeks later at a place that allegedly tallied with the envisioned Wyoming site (Graff 1998, 14; Schnabel 1997, 342–43).

Some sources do not give the psychic's name, but she was in fact Angela Dellafiora, a former civilian analyst for the Army's Intelligence and Security Command (INSCOM) in Latin America. Seemingly dissatisfied with that work, she had been increasingly drawn to the mystical and joined the remote-viewing unit in early 1986. Although she had tried the psychic sensing technique employed by the others in the unit, Dellafiora soon found she obtained better results by relying on alleged spirit communication. She would lapse into a "trance," whereupon one of her entities—known as "George," "Dr. Einstein," and, most often, "Maurice"—would possess her body and manipulate her writing hand to produce responses to questions posed by her monitor.

Thus, Dellafiora was not actually practicing remote viewing as it is usually understood. Although her automatic writing technique came to be called "written RV," Jim Schnabel correctly observed that "it was essentially a form of spirit mediumship—in modern parlance, 'channeling'" (1997, 342). The males in the program were unhappy with the involvement of "spirit guides" (Morehouse 1996, 128), referring disparagingly to Dellafiora's spirit entities as "the boys" or even "The Three Stooges." They saw the unit regressing from "high-tech wizardry back to archaic and vaguely feminist witchery" (Schnabel 1997, 343–44). To skeptics, their attitude may seem a case of the pots calling the cauldron black.

Some thought Dellafiora had "achieved an undue influence on the unit when she began to give personal "channeling sessions" to some of the DIA officials, "featuring advice on the most intimate matters of their lives" (Schnabel 1995, 13). One of the other psychic viewers, Mel Riley, groused, "They were told all the nice things they wanted to hear, which reinforced Angela's position within the unit" (Schnabel 1995, 13).

More to the point, most of the other remote viewers reportedly thought Dellafiora "was prone to wild errors." For example, in the case of Lieutenant Colonel William Higgins, who was kidnapped by Islamic terrorists in 1988, Dellafiora envisioned him alive, believed he was being held in an underground location, and reported that he was soon to be released. In fact, he had probably been kept in a Lebanese house, and before long his tortured corpse was recovered.

Moreover, even when Dellafiora's channeling seemed successful, her remote-viewing colleagues suggested that sometimes the results were not entirely due to paranormal ability. They felt she "was too often inadvertently coached towards targets by her customers' questions and answers" (Schnabel 1997, 345). By asking for and obtaining feedback, which enabled her to correct her course, she was naturally able to more accurately describe a given target.

"Seeing" Charlie Jordan

Such criticisms have serious implications for the Charlie Jordan matter. Accounts of Angela Dellafiora's touted success in that case typically fail to mention such problems and, indeed, offer conflicting claims about what really happened. Most versions seem to stem from the reports of Dale E. Graff, the originator and former project director of Stargate and author of *Tracks in the Psychic Wilderness*. Obviously referring to the Jordan case, Graff wrote (1998, 14):

> We were asked to locate a former Drug Enforcement Agency (DEA) employee who was a fugitive wanted for drug-smuggling cooperation. One of the Stargate remote viewers narrowed down his location to northern Wyoming near a campground. Although our data was not acted upon, the fugitive was captured a few weeks later at a campground in northern Wyoming. This data was totally contrary to the DEA expectations. They believed he was hiding somewhere in the Caribbean region.

Later, Graff (2000, 106) wrote that the psychic (Dellafiora) had said of Jordan, "He is in Wyoming, near a place that sounds like Lowell. There is an Indian burial place nearby." Note that in this version she did not state *northern* Wyoming. There is in fact no Lowell in that entire state, but when a town named Lovell was found on a map, Dellafiora was then "sure that he was somewhere in the Northwest part of Wyoming, even if not exactly in Lovell." (Other accounts say that the psychic reported "Low, Wyoming" [Topping 1999] or "Low, Low, Lowell? Lowell. Lowell, Wyoming. It's Wyoming" [*Mysteries* 1998].)

Citing several Stargate sources, including Graff, Jim Schnabel states that Dellafiora located the fugitive "in northern Wyoming, near the town of Lovell and also not far from an old Indian burial ground." Because of "conflicting information" from the other psychics in the unit, the Customs Service ignored all of the pronouncements, but later "Jordan was spotted by a ranger at Yellowstone National Park—a few dozen miles from Lovell, Wyoming—and was arrested. Under interrogation, Jordan admitted to having been near Lovell around the time Angela had psychically placed him there" (Schnabel 1997, 342–43).

Still another source, the *Washington Post*, reported that "Jordan was found in Pinedale, Wyo., not far from Yellowstone National Park—in a campground near an Indian burial site" (Anderson and Moller 1996). Note that the Indian burying ground motif has been transposed from near Lovell (actually the nonexistent Lowell)—where Jordan supposedly was when psychically spotted—to Pinedale, where he was captured. (Pinedale is nearly 100 aerial miles south of Yellowstone, or about 135 driving miles.)

One source cited an unnamed "former customs official" as confirming that "[t]he work of the psychic was 'instrumental' in Jordan's capture" (Anderson and Moller 1996). This claim is belied by Stargate project manager Dale Graff's previously quoted statement (1998, 14) that "our data was not acted on" and by investigative writer Schnabel's report (1997, 342) that "[t]he Customs Service decided to ignore" the wealth of "conflicting information" provided by the remote viewers.

The unnamed retired customs officer was almost certainly Bill Green, who at the time of the Jordan manhunt was Assistant Commissioner of Internal Affairs for the Customs Service. He did tell the BBC, "I made sure that the police in Wyoming were made aware of the possibility that Charlie could be in their state" (*Mysteries* 1998). However, some time before Jordan's capture, the police had independently spotted his vehicle outside of Denver, Colorado (from whence Interstate 25 leads to the highway, alternate U.S. 14, that runs through Lovell, Wyoming). In short, authorities may already have been alerted to Wyoming as a possible area to search for the fugitive.

Yet another permutation of the proliferating tale was summed up by the narrator for the BBC *Mysteries* program. Although based on interviews with the principals—notably Stargate Project Manager Dale Graff and former Customs officer Bill Green—this version is incorrect in every detail: "The arrest was made in northwest Wyoming [in fact Pinedale is in *southwest* Wyoming, although just outside the northwest quadrant], a hundred miles from Lovell [actually 300 driving miles from there, or about 160 miles as the crow flies], next to an Indian burial ground [though not according to Graff's account in his *River Dreams*, as we have seen]."

Obviously the details of the case are now badly garbled, and it is difficult to say what, if anything, Angela Dellafiora should be given credit for. One skeptic who appeared on the BBC program with me, psychologist Chris French, commented:

> If we accept this case at face value, then it might actually appear very impressive; but, as a scientist, I never accept these kinds of cases at face value. The kind of questions we would need to look at further would be how much information did she have access to, bearing in mind that this particular case had been featured on the TV program in America, *America's Most Wanted*" [*Mysteries* 1998].

In fact, it now becomes apparent that there is no real way to know—no official, detailed record to specify—just what information was supplied to the various remote viewers, including Dellafiora, or precisely what "psychic" information or predictions they provided, let alone dependable information about the alleged "hits" (e.g., the Indian burial site). Referring to archived records in the matter, which have never been published or cited as documentation, Graff conceded that "some of the pieces are not in the files." He did add, rather lamely, that "I have some in my journals so it's not totally relying on memory" (*Mysteries* 2000).

It would be nice to know, for instance, whether Dellafiora made other—perhaps completely inaccurate—pronouncements in the Charlie Jordan case that have since been conveniently forgotten. Alleged psychic sleuths typically depend for their apparent successes on a technique called *retrofitting*. This involves giving out several vague "clues" and then—when the case is solved, usually through conventional police work—doing after-the-fact matching so the reputed clairvoyant can take credit for the success. Errors are ignored or rationalized away, and dubious pronouncements are interpreted as necessary to make "hits." For example, "water" can be interpreted as indicating a nearby creek, river, or other body of water; a place name, such as Riverside Drive or Lakeshore Boulevard; a structure such as a water tower or hydroelectric plant; or some other possibility (Nickell 1994, 15–16, 182). Credulous folk may inadvertently help the mystic through selective memory, creative interpretation, exaggeration, and other processes.

Not only may one pick and choose among the pronouncements of a single psychic, but a similar selection process may also be applied to the psychics themselves. In the Jordan case, several remote viewers undertook 18 sessions, all logged in 1989, in the attempt to locate Jordan (Anderson and Moller 1996). All of these were apparently worse than useless, except for the alleged offerings of Dellafiora. As psychic investigator Milbourne Christopher wryly commented in his *ESP, Seers & Psychics* (1970, 81), "Fire enough shots, riflemen agree, and eventually you'll hit the bull's-eye."

In summary, the Charlie Jordan case, touted as one of the most successful examples of remote viewing from the U.S. government's psychic-spying project, is not convincing evidence of anything—save perhaps folly. Not only was the case actually an example of alleged spirit contact rather than extrasensory perception, but it also illustrates the limitations of anecdotal evidence: conflicting versions, selective reporting, and lack of documentation, together with additional manifestations of faulty memory, bias, and other human foibles.

The evaluators of the intelligence-gathering usefulness of remote viewing concluded that the technique "has not been shown to have value in intelligence operations." They found that the information provided was vague, inconsistent (from viewing to viewing), and often irrelevant or outright erroneous. As mentioned earlier, they also stated that in some cases of touted hits, the remote viewers might have had much more background information than was apparent. They determined that "remote viewings have never provided an adequate basis for 'actionable' intelligence operations—that is, information sufficiently valuable or compelling so that action was taken as a result" (Mumford et al. 1995). The Charlie Jordan case seems to have been no exception.

REFERENCES

Anderson, Jack, and Jan Moller. 1996. Military psychic unit's "hits" and misses. *Washington Post*, 30 December.

Christopher, Milbourne. 1970. *ESP, Seers & Psychics*. New York: Thomas Y. Crowell.

Graff, Dale E. 1998. *Tracks in the Psychic Wilderness*. Boston: Element.

———. 2000. *River Dreams: The Case of the Missing General and Other Adventures in Psychic Research*. Boston: Element.

Guiley, Rosemary Ellen. 1991. *Harper's Encyclopedia of Mystical & Paranormal Experience*. New York: HarperCollins.

Hyman, Ray. 1996. Evaluation of the military's twenty-year program on psychic spying. *Skeptical Inquirer* 20, no. 2 (March/April): 21–26.

Lamothe, John D. 1972. Controlled offensive behavior—USSR. Report for Defense Intelligence Agency.

Morehouse, David. 1996. *Psychic Warrior: Inside the CIA's Stargate Program*. New York: St. Martin's Press.

Mumford, Michael D., et al. 1995. An evaluation of remote viewing: Research and applications. Prepared for the Central Intelligence Agency by The American Institutes for Research, 29 September.

Mysteries. 1998. "The Charlie Jordan Case" segment. BBC, aired 23 November.

Nickell, Joe, ed. 1994. *Psychic Sleuths: ESP and Sensational Cases*. Buffalo, N.Y.: Prometheus Books.

Ostrander, Sheila, and Lynn Schroeder. 1971. *Psychic Discoveries Behind the Iron Curtain.* New York: Bantam.

Schnabel, Jim. 1995. Tinker, tailor, soldier, psi. *Independent on Sunday* (London), 27 August, 10–11, 13.

———. 1997. *Remote Viewers: The Secret History of America's Psychic Spies.* New York: Dell.

Topping, Jane. 1999. Personal communication to Joe Nickell, 24 May, of BBC research notes following telephone interviews with Dale Graff and retired Customs official Bill Green.

10

Amityville
The Horror of It All

The best-selling book *The Amityville Horror: A True Story* (Anson 1977) was followed by a movie of the same title and a sequel, *Amityville II: The Possession*. Although the original proved to be a hoax, that fact does not seem well known to the general public. A book published in 2002 now sheds new light on the sordid affair and reviews the multiple-murder case that preceded it. Written by Ric Osuna, it is titled *The Night the DeFeos Died: Reinvestigating the Amityville Murders*.

The saga began on November 13, 1974, with the murders of Ronald DeFeo Sr.; his wife, Louise; and their two sons and two daughters. The six were shot while they slept in their home in Amityville, New York, a community on Long Island. Subsequently the sole remaining family member—Ronald Jr., nicknamed "Butch"—confessed to the slaughter and was sentenced to 25 years to life in prison. Just two weeks after his sentencing, late the following year, George and Kathy Lutz and their three children moved into the former DeFeo home where—allegedly—a new round of horrors began.

The six-bedroom Dutch Colonial house was to be the Lutzes' residence for only 28 days. They claimed they were driven out by sinister forces that ripped open a heavy door, leaving it hanging from one hinge; threw open windows, bending their locks; caused green slime to ooze from a ceiling; peered into the house at night with red eyes and left cloven-hoofed tracks in the snow outside; infested a room in mid-winter with hundreds of houseflies; and produced myriad other supposedly paranormal phenomena, including inflicting inexplicable, painful blisters on an investigating priest's hands.

Local New York television's Channel 5 "investigated" the alleged haunting by bringing in alleged psychics, together with "demonologist" Ed Warren and his wife, Lorraine, a professed "clairvoyant." The group held a series of séances in the house. One psychic claimed to have been made ill and to "feel personally threatened" by shadowy forces. Lorraine Warren pronounced the presence of a negative entity "right from the bowels of the earth." A further séance was unproductive, but the psychics agreed that a "demonic spirit" had possessed the house and recommended exorcism (Nickell 1995, 122–29).

In September 1977, the book called *The Amityville Horror: A True Story* appeared. Written by Jay Anson, a professional writer commissioned by Prentice-Hall to tell the Lutzes' story, it became a runaway best-seller. Anson asserted: "There is simply too much independent corroboration of their narrative to support the speculation that they either imagined or fabricated these events," although he conceded that the strange occurrences ceased after the Lutzes moved out.

Indeed, a man who later lived in the house for eight months said he had experienced nothing more horrible than a stream of gawkers who tramped onto the property. Similarly, the couple who purchased the house after it was given up by the Lutzes, James and Barbara Cromarty, poured ice water on the hellish tale. They confirmed the suspicions of various investigators that the whole story had been a bogus admixture of phenomena: part traditional haunting, part poltergeist disturbance, and part demonic possession, including elements that seemed to have been lifted from the movie *The Exorcist*.

Researchers Rick Moran and Peter Jordan (1978) discovered that the police had never been called to the house, and that there had been no snowfall when the Lutzes claimed to have discovered cloven hoofprints in the snow. Other claims were similarly disproved (Kaplan and Kaplan 1995).

I talked with Barbara Cromarty on three occasions, including when I visited Amityville as a consultant to the *In Search Of* television series. She told me not only that her family had experienced no supernatural occurrences in the house, but also that she had evidence showing that the whole affair was a hoax. Subsequently, I recommended to a producer of the then-forthcoming TV series *That's Incredible*, who had asked my advice about filming inside the house, that they have Mrs. Cromarty point out various discrepancies for close-up viewing. For example, recalling the extensive damage to doors and windows detailed by the Lutzes, she noted that the old hardware—hinges, locks, doorknob, and the like—were still in place. Upon close inspection, one could see that there were no disturbances in the original paint and varnish (Nickell 1995).

In time, Ronald DeFeo's attorney, William Weber, told how the Lutzes had come to him after leaving the house, whereupon he had told them that their "experiences" could be useful to him in preparing a book. "We created this horror story over many bottles of wine that George Lutz was drinking," Weber told the Associated Press. "We were creating something the public wanted to hear about." Weber later filed a $2-million suit against the couple, charging them with reneging on their book deal. The Cromartys also sued the Lutzes, Anson, and the publishers, maintaining that the fraudulent haunting claims had resulted in sightseers destroying any privacy they might have had. During the trials, the Lutzes admitted that virtually everything in *The Amityville Horror* was pure fiction (Nickell 1995, Kaplan and Kaplan 1995).

Nevertheless, an astonishingly biased treatment of the case aired on ABC News' *Primetime* on Halloween 2002. TV personality Elizabeth Vargas seemed bent on believing the discredited George Lutz and attacking—or ignoring—skeptics. My comments mostly ended up on the cutting-room floor. In one lost moment, I stated that the police had not been called to the house as the book claimed. Vargas—who was hostile to all my evidence—retorted that I did not know what I was talking about and that such a claim was not in the book, which she insisted she had just read. In fact, it is on page 157 of the original paperback edition (Anson 1977). (For more on Vargas's pathetically credulous treatment of Amityville, see Christopher 2002.)

Now Ric Osuna's *The Night the DeFeos Died* adds to the evidence. Ronald DeFeo's wife, Geraldine, allegedly confirmed much of Weber's account. To her, it was clear that the hoax had been planned for some time. Weber had intended to use the haunting claims to help obtain a new trial for his client (Osuna 2002, 282–86). As to George Lutz—now divorced from his wife and criticized by his former stepsons—Osuna states that "George informed me that setting the record straight was not as important as making money off fictional sequels." Osuna details numerous contradictions in the story of which Lutz continues to offer various versions (2002, 286–89).

For his part, Osuna has his own story to tell. He buys Ronald "Butch" DeFeo's current story about the murders, assuring his readers that it "is true and has never been made public" (2002, 18, 22). DeFeo now alleges that his sister Dawn urged him to kill the entire family and that she and two of Butch's friends also participated in the crimes. In fact, Butch maintains that Dawn began the carnage by shooting their domineering father with the .35-caliber Marlin rifle. Butch then shot his mother, whom he felt would have turned him in for the crime, but claims he never intended to kill his siblings. He left the house to look for one of his friends who had left the scene; when he returned to find that Dawn had murdered her sister and other two brothers, he was enraged. He fought with her for the gun and sent her flying into a bedpost where she was knocked out. He then shot her.

Osuna tries to make this admittedly "incredible" tale believable by explaining away contradictory evidence. Osuna accepts DeFeo's claim that he altered the crime scene, and asserts that the authorities engaged in abuses and distortions of evidence to support their theory of the crimes. Even so, Osuna concedes that "Butch had offered several different, if ludicrous, versions of what had occurred" (2002, 33), and that he might again change his story. Nevertheless, Osuna asserts that "[t]oo much independent corroboration exists to believe it was just another one of his lies" (370).

I remain unconvinced. Butch DeFeo has forfeited his right to be believed, and his current tale is full of implausibilities and contradictions. Osuna appears to me simply to have become yet another of DeFeo's victims.

REFERENCES

Anson, Jay. 1977. *The Amityville Horror: A True Story.* New York: Bantam Books.

Christopher, Kevin. 2002. The ABC-ville horror. *Skeptical Inquirer* 27, no. 1 (January/February): 53–54.

Kaplan, Stephen, and Roxanne Salch Kaplan. 1995. *The Amityville Horror Conspiracy.* Lacyville, Pa.: Belfrey Books.

Moran, Rick, and Peter Jordan. 1978. The Amityville Horror hoax. *Fate* (May): 44–46.

Nickell, Joe. 1995. *Entities: Angels, Spirits, Demons, and Other Alien Beings.* Amherst, N.Y.: Prometheus Books.

Osuna, Ric. 2002. *The Night the DeFeos Died: Reinvestigating the Amityville Murders.* N.p.: Xlibris.

11

Sideshow!

Investigating Carnival Oddities and Illusions

Like Robert Ripley, I have always been attracted to the odd and the curious. Growing up in a small town, I tried never to miss a visiting solo act—like an armless wonder or a bullwhip artist—who performed at the local ball park. I paid admission to countless magic, hypnotism, and spook shows, not to mention animal and juggling acts, that played in the school auditorium or the local theater. And I must have attended every carnival and circus that came around.

In 1969 I worked as a magic pitchman in the carnival at the Canadian National Exhibition. It was there that I met "El Hoppo the Living Frog Boy" and witnessed the transformation of "Atasha the Gorilla Girl," who changed from beauty to beast before the eyes of frightened spectators (Nickell, 1970). During travels in Europe, Asia, and North Africa in 1970 and 1971, I beheld various street acts, including nighttime fire-breathing and Houdini-style chain-escape performances in

Paris, a "dancing" bear in Istanbul, a little old wandering conjurer at the Pueblo Español in Barcelona, and a snake charmer and other entertainers at the Medina in Marrakech.

Barnum and Sideshows

Such street performances and performers hark back to the earliest form of what developed into the great English fairs of the early Renaissance. There, most of the "human curiosities" that later became fixtures of nineteenth-century American "freak shows" were exhibited (Bogdan 1990, 25). In late 1841, an itinerant showman named P. T. Barnum became the proprietor of the American Museum in New York City, an entertainment enterprise that had featured contortionists, a banjoist, a lady magician, a lecturer on animal magnetism, a Tattooed Man, and similar acts (Harris 1973, 40).

Barnum had earlier toured with Joice Heth, supposedly the 161-year-old former nurse of George Washington but actually an octogenarian fraud. After taking over the American Museum, he exhibited the "Feejee Mermaid," billed as "the greatest Curiosity in the World" although it was only a monkey's body grafted onto a fish's tail (Harris 1973, 22, 62–67). Accusations of trickery only brought Barnum increased notoriety, and he soon schemed to have his bearded lady accused of being a man! A well-publicized medical examination helped boost cash receipts. When one visitor asked whether an exhibit was real or a humbug, Barnum replied, "That's just the question: persons who pay their money at the door have a right to form their own opinions after they have got up stairs" (Harris 1973, 77).

Barnum exhibited increasingly diverse oddities, such as albinos, giants, dwarfs, and "The Highland Fat Boys," along with ballets, dramas, magic shows, and "scientific demonstrations." By the 1870s, dime museums (Barnum's was 25 cents) began to proliferate, and "the human oddity was the king of museum entertainment" (Bogdan 1990, 32–33, 37). Traveling museums, linked to circuses as concessions, presaged the later "sideshows"—so named because they were separate from the main attraction.

Actually, a circus could have several sideshows, located in tents (or later trailers) on the *midway*, the place where the rides, shows, games, and refreshments are located. A carnival is essentially *only* a midway (Taylor 1997, 92–95).

The Ten-in-One

A major type of sideshow, often popularly called a *freak show* because human oddities were usually among the exhibits, was known to insider "carnys" as a "ten-in-one." As its name indicated, it consisted of a number of acts, often arrayed along a platform, with the crowd moving from one to the other in sequence. Because such shows were typically continuous, if a spectator entered the tent during, say, the sword swallower's performance, he or she would be led by the "lecturer" through the remaining nine (approximately) acts or features—magician, fat lady, giant, and so on. When the sword swallower was on again, that was the signal to exit the show.

At the end of each act or exhibit, spectators might be offered a "pitched" item to buy, such as a "true life" booklet or photograph. Frequently giants sold huge finger rings and midgets offered miniature Bibles. (I bought an autographed photo from "El Hoppo the Living Frog Boy" and an envelope of tricks from a magician.) Such an extra, inside sale is known as an "aftercatch" (Taylor 1997, 91). (See Figure 11-1.)

Meanwhile, outside, a "talker" (real carnys never use the term *barker*) was periodically drumming up a new crowd (or *tip*) of potential customers, usually with the assistance of one or more of the acts to provide a taste of what was inside. This external pitch was held on a "bally" platform, the name deriving from *ballyhoo* (meaning sensationalized promotion).

The oddities and exotic acts that were featured in ten-in-one shows were quite varied. In his book *Monster Midway* (1953, 102), Wil-

FIGURE 11-1. Pitch card of a sideshow snake charmer (author's collection).

FIGURE 11-2. *Carte de visite* picture of midget Tom Thumb's 1863 wedding, promoted by P. T. Barnum (author's collection).

liam Lindsay Gresham discussed the traditional carny classification of human oddities, observing that "[i]n addition to a born, bona fide freak, the same show will sometimes feature 'made' freaks and 'gaffed' [fake] freaks, all scrambled together." For the following discussion, I have subdivided the first category and added other nonoddity divisions in an attempt to provide a more complete classification of sideshow acts and exhibits. (Sideshow attractions like the Fun House are not included.)

Oddities

One may think of "born" human oddities as of essentially two types. First is the more-or-less obvious anomaly. Examples include midgets like Barnum's "Tom Thumb" (Charles Sherwood Stratton) and Lavinia Warren, who married in a highly promoted ceremony (Drimmer 1991, 172–82). (See Figure 11-2.) At the other end of that spectrum was Jack Earle, whose extreme height got him noticed by a Ringling Brothers circus sideshow manager in the mid 1920s. "How would you like to be a giant?" the showman is said to have asked, indicating the important distinction between being merely noticeable and being a sideshow star. Earle soon became "The Texas Giant" (Bogdan 1990, 280).

Another example of the true type of oddity is conjoined twins, the result of incomplete separation of a single, fertilized egg. The most celebrated pair were Chang and Eng (1811–1874) who came from Siam and thus begat the term "Siamese twins." They each eventually married, living in three-day shifts in their respective houses and fathering 21 children (Drimmer 1991, 3–27).

Sometimes the division of the single, fertilized egg that produces identical twins is even less complete than it was with Chang and Eng. The result can be any of varkious anatomical oddities, such as "The Two-Headed Boy"—actually the Tocci brothers (b. 1877), who were two individuals above the sixth rib but who shared a single body below. In some cases the incomplete division results in a normal-size body with a smaller, parasitic one—in whole or part—connected to it. Such was the case with "The Four-Legged Girl from Texas" (Myrtle Corbin), "The Man with Two Bodies" (Jean Libbera, b. 1884), and "The Girl with Four Legs and Three Arms" (Betty Lou Williams, d. 1955) (Drimmer 1991, 28–37; Parker 1997, 64).

Other genuine oddities include hirsute people like bearded ladies and "Lionel the Lion-faced Man," whose face was entirely covered with long hair (Parker 1997, 92, 94), as well as various "Alligator Boys and Girls" afflicted with the skin condition ichthyosis. Still others, like "Leona the Leopard Girl," were dark-skinned people with vitiligo, a lack of pigmentation that could appear as a pattern of white splotches over the body (Meah 1996, 120–22).

Yet another example of the genuine anomaly was the Frog Boy, although any of various deformities could qualify one for the sobriquet. There was "El Hoppo" (previously mentioned, whom I met in 1969). Although the sideshow banner depicted a youth with a frog's hindquarters, in actuality "Hoppy" was a grey-bearded man in a wheelchair, who had spindly limbs and a distended stomach. To make himself look more froglike, he wore green leotards (Nickell 1995, 221–22). Among others, there was Otis Jordan, an African-American who had (according to one of his many admirers) the body of a four-year-old but a normal head with "a noble, scholarly face" (Meah 1998, 56). Beginning in 1963, he performed as "Otis the Frog Boy"; part of his routine was to roll, light, and smoke a cigarette using only his lips. When his act was shut down in 1984, after a woman complained about the exhibition of disabled persons, Otis moved to Coney Island, where he continued his act with the more politically correct billing, "The Human Cigarette Factory" (Bogdan 1990, 1, 279–81; Taylor 1998, 55–61).

As with human "frogs," other examples of genuine anomalies were also imaginatively interpreted: "The Caterpillar Man," also known as "Prince Randian, the Hindu Living Torso"; "The Mule-Faced Woman," Grace McDaniels, who had facial tumors; various persons who had vestigial feet and hands attached to the torso, such as "Sealo the Seal Boy" and "Dickie the Penguin Boy" (who, his banner proclaimed, "Looks and Walks Like a Penguin"); and many others (Fiedler 1993, 23, 168–70, 291; Johnson, Secreto, and Varndell 1996, 68, 126; Taylor 1997, 95).

A second subclass of the "born" oddity is what is known in carny parlance as the "anatomical wonder," that is, "a sideshow performer, usually perceived as a human oddity, but more a working act" (Taylor 1997, 91). A good example is James Morris, who performed with Barnum and Bailey for many years. He could stretch the skin of his cheek eight inches and pull his chest skin to the crown of his head. Morris was only one of many who were styled "The Elastic Skin Man" (or Woman). Others who had the same harmless condition, known as *cutis hyperelastica*, were billed as "The India Rubber Man" or similar designation (although that term probably more often referred to a contortionist) (Drimmer 1991, 307; Taylor 1998, 95). Other anatomical wonders included "Popeye, the Man with the Elastic Eyeballs," who could cause one or both of his eyes to protrude to an incredible degree. Charles Tripp, "The Armless Wonder," teamed up with Eli Bowen, "The Legless Wonder," to perform amusing stunts like riding a bicycle built for two (Drimmer 1991, 87–93; Johnson, Secreto, and Varndell 1996, 48).

The second main category of oddities—what Gresham termed "'made' freaks"—is typified by tattooed persons. That sideshow genre was popularized after a Russian explorer's visit to the Marquesas Islands in 1804. He discovered a French deserter named Jean Baptiste Cabri who had married a native woman and been extensively tattooed. Cabri returned with the explorer to Moscow, where he launched a theatrical career, then toured Europe, regaling audiences with exaggerated tales (Johnson, Secreto, and Varndell 1996, 101–2).

Probably the most unique of the tattooed men and women (both eventually appeared on sideshow banners) was Horace Ridler, a British prep-school-educated ex-army officer who was down on his luck and decided to transform himself into a circus star. His idea was to become tattooed all over with zebra-like stripes—a process that took a year beginning in 1927. Claiming that he had been forcibly tattooed by New Guinea savages, "The Great Omi, The Zebra Man," eventually became "one of the highest paid circus performers in the world" (Gilbert 1996, 104; Bogdan 1990, 255–56).

Other "made" freaks include a "crucified man," Mortado, who had had his hands and feet pierced surgically. In the holes he concealed capsules of "blood" that spouted forth when spikes were pounded through them. Later, using a specially designed chair with plumbing fixtures, he became Coney Island's "Mortado the Human Fountain."

Then there were the "gaffed"—faked—freaks. Such manufactured oddities included phony Siamese twins like Adolph and Rudolph. A circa-1899 photograph reveals that they lacked the close resemblance of identical twins (which conjoined persons always are). In fact, a harness concealed under their specially devised suit held Rudolph so that he seemed to grow from Adolph's waist (Bogdan 1990, 8; Reese 1996, 190). Fake "Alligator" girls and boys were created by painting their bodies with a weak solution of glue; after allowing it to dry, they twisted and flexed to create a cracking effect that simulated ichthyosis (Meah 1996, 120).

Sometimes gaffing was done to enhance a true oddity. A good example was William Durks, whose deformity led him to be billed as "The Man with Two Faces" (among other appellations). In addition to a cleft palate, Durks had an eye and nostril on either side of a growth in the center of his face. He later enhanced the effect by using makeup to add an extra central "eye" and two "nostrils," becoming "The Man with Three Eyes." Actually Durks was *one*-eyed, his other being vestigial (Taylor 1997, 40–47).

In packaging their exhibits, showmen typically exaggerated claims and fabricated backgrounds. For example, dwarfs and midgets had inches subtracted from their height, and giants often wore lifts and tall hats to enhance theirs, which was inflated by as much as 12 inches (Bogdan 1990, 95–97).

Wonder-Workers

After human oddities, the second major category of sideshow performers consists of those who exhibit a special skill. They include sword swallowers, who must learn to conquer the gag reflex in order to swallow not only swords—like Edith Clifford (b. 1884), "Champion Sword Swallower of the World"—but also umbrellas and lit neon tubes (Houdini 1920, 147–51; Mannix 1951, 96–101).

Other performers in this class are the fire eaters and fire breathers (who sip flammable liquid and spew it across a torch to produce great fireballs). Then there are performers of various "torture" acts: the Human Pincushion (who sticks needles through the flesh); the Human Blockhead (who hammers spikes up the nose); and others, including "fakirs" who lie on beds of nails. Other wonder-workers are snake charmers (whose act might consist of little more than wrapping a large snake about the body [again see Figure 11-1]); contortionists like "Huey the Pretzel Man"; and numerous "Strong Men" and women, including Louis Cyr, whom Houdini (1920, 221) suggested was "the strongest man in the known world at all-around straight lifting." In this strongman subcategory were William Le Roy (b. 1873), "The Human Claw-Hammer," who could extract a nail driven through a two-inch plank using only his teeth; and Madame Rice, "The Most Diminutive Lady Samson in the World" (Taylor 1997, 91–96; Johnson, Secreto, and Varndell 1996, 78; Houdini 1920, 223–24; Bogdan 1990, 265).

Illusions

A third major class of sideshow features is represented by what is known as an "illusion show." An example—as old as it is effective—is a transformation effect such as girl-to-gorilla, skeletal-corpse-to-living-vampire, and so on (Taylor 1997, 93, 94). In 1969, on a break from my stint as a carnival pitchman, I joined spectators in a sideshow tent to see "Atasha the Gorilla Girl" standing, apparently, at the rear of a cage. As a voice chanted, "Goreelyagoreelyagoreelya, ATASHA, goreelya!" Atasha's features were slowly transformed into those of a large gorilla. Suddenly, it rushed from the unlocked(!) cage and lunged toward the crowd, sending some spectators screaming from the exit—an occurrence that helped draw the next tip (Nickell 1970, Teller 1997).

Of course, the effect was a magician's trick. Often the bally talker slyly noted that the "Gorilla Girl"—or the "victim" in another illusion termed "The Headless Woman"—was in "a legerdemain condition." Other illusions commonly featured in sideshows were "The Girl in the Fish Bowl" (wherein a living "mermaid" appears in apparent miniature in a goldfish bowl) and "Spidora, the Spider Girl" (which consisted of a living human head atop an arachnid's body) (Taylor 1997, 21, 93, 94).

An illusion of early vintage that was especially popular around the end of the nineteenth century was an effect known to magicians and carnys as a "blade box." A young woman would lie in a box that was then intersected with a number of blades (Figure 11-3). The secret? To learn that, one paid an extra charge (another form of aftercatch called a *ding*) to come up on the platform and peer inside the box. To provide extra incentive to the male spectators, the magician might reach in and pull out his assistant's costume! The spectators were thus fooled twice, since the costume was an extra one ("Science" 1997; Taylor 1997, 92).

Animals

Still another major type of sideshow exhibit features animals. While the premier acts are shown under the circus Big Top, midways and carnivals often have sideshow animal presentations. In 1972 in Toronto, I visited an all-animal ten-in-one. It included a three-legged sheep, touted as "Nature's Living Tripod," and various alleged hybrids (zebra/donkey, turkey/chicken, dog/raccoon). These did not match their banner portraits, which showed the front half of one attached to the rear of another, but merely resembled a blend of features. There was also a ram with four horns, a sheep and a cow with five legs each, and other oddities.

FIGURE 11-3. **Carny showman Bobby Reynolds presents a blade-box illusion at New York's Erie County Fair, 1999. (Photograph by the author.)**

As billed, the "World's Smallest Horse" was a "preserved exhibit" (a fetus pickled in a jar!), and the "World's Largest Horse" was indeed in "photographic form." To distinguish the living exhibits from such "curios" (as I describe them in the next section), banners still typically feature the screaming word "ALIVE."

With the decline of the ten-in-one in the 1980s—due to their high overhead and the fact that the exhibition of human oddities could provoke complaints—individual animal and illusion exhibits became the mainstay. One was the "Giant Rat" show that I witnessed at the Kentucky State Fair (see Figure 11-4). In such exhibits the giant creature was either of two types of South American aquatic rodents, usually the capybara (which belongs to the guinea pig family) (Taylor 1997, 20, 93; *Encyclopædia Britannica* 1960).

FIGURE 11-4. **"Giant Rat," an individual sideshow feature at many carnivals. Note the word "ALIVE." (Photograph by the author.)**

Curios

A fifth and last category of sideshow exhibits is reserved for any inanimate object, including preserved human or animal specimens. Barnum's "Feejee mermaid" is one (albeit gaffed) example. Another is any of the various sideshow mummies, such as one alleged to be of John Wilkes Booth that was exhibited throughout the first half of the twentieth century (Quigley 1998, 69).

Curios I have paid admission to see include the bullet-riddled car of outlaws Bonnie and Clyde; a "sasquatch" (actually a rubber fake) "safely frozen in ice" (Nickell 1995, 230); and a concrete copy of the famous hoaxed petrified man that was billed as "the Cardiff Giant, ten feet four inches." Although the fine print on the bottom of the banner confessed, "This is a facsimile," the talker promised, "He's a big son of a gun!"

Exit This Way

Most ten-in-ones featured an extra attraction (or *blowoff*), typically curtained from view, that functioned like an aftercatch to the entire show. For an extra fee, one might see a five-legged horse or an illusion like the Headless Woman (Mannix 1951, 45; Bogdan 1990, 103–4).

Often a spectator would ask of an exhibit, "Is it real?" Showman Ward Hall responded for carnys everywhere: "Oh, it's *all* real. Some of it's really real, some of it's really fake, but it's all really good" (Taylor 1997, 81). Echoing the sentiment was legendary showman Bobby Reynolds, whose traveling "International Circus Sideshow Museum & Gallery" featured a huge banner ballyhooing "The Really Real Frog Band! Real Frogs!" Outfitted with miniature clarinets, drums, and other instruments is a band of stuffed amphibians. Did Reynolds get any complaints from the tip? "No. They'd look at it, they'd say, 'Do these frogs play?' and I'd say, 'Well, they used to.' 'Are they real frogs?' 'They're real frogs.' 'Why don't they play?' 'They're dead'" (Taylor 1997, 22–23).

Carnys developed an us-versus-them attitude that derived from the hostility they frequently encountered from "rubes" (the locals). In the carnival subculture, outsiders could be targets for rigged games, shortchanged ticket sales, and other scams (Bogdan 1990, 88–89). For those forewarned—like readers of this introduction to sideshows—there was, and is, much to learn from and appreciate in carnivals.

REFERENCES

Bogdan, Robert. 1990. *Freak Show: Presenting Human Oddities for Amusement and Profit*. Chicago: University of Chicago Press.

Drimmer, Frederick. 1991. *Very Special People*. New York: Citadel Press.

Encyclopædia Britannica. 1960. s.v. "capybara."

Fiedler, Leslie. 1993. *Freaks: Myths and Images of the Secret Self*. New York: Doubleday.

Gilbert, Steve. 1996. Totally tattooed. In *Freaks, Geeks & Strange Girls: Sideshow Banners of the Great American Midway,* edited by Randy Johnson, Jim Secreto, and Teddy Varndell, 101–5. Honolulu: Hardy Marks Publications.

Gresham, William Lindsay. 1953. *Monster Midway.* New York: Rinehart.

Harris, Neil. 1973. *Humbug: The Art of P. T. Barnum.* Chicago: University of Chicago Press.

Houdini, Harry. [1920] N.d. *Miracle Mongers and Their Methods.* Reprinted Toronto: Coles.

Johnson, Randy, Jim Secreto, and Teddy Varndell. 1996. *Freaks, Geeks & Strange Girls: Sideshow Banners of the Great American Midway.* Honolulu: Hardy Marks Publications.

Mannix, Dan. 1951. *Step Right Up!* New York: Harper & Brothers.

Meah, Johnny. 1996. Notes on alligator skinned people. In *Freaks, Geeks & Strange Girls: Sideshow Banners of the Great American Midway,* edited by Randy Johnson, Jim Secreto, and Teddy Varndell, 120–25. Honolulu: Hardy Marks Publications.

———. 1998. The Frog Prince. In *James Taylor's Shocked and Amazed! On and Off the Midway,* vol. 5, 54–61. Baltimore: Dolphin-Moon Press/Atomic Books.

Nickell, Joe. 1970. Magic in the carnival. *Performing Arts in Canada* 7, no. 2 (May): 41–42.

———. 1995. *Entities.* Amherst, N.Y.: Prometheus Books.

Parker, Mike. 1997. *The World's Most Fantastic Freaks.* London: Hamlyn.

Quigley, Christine. 1998. Mummy Dearest. In *James Taylor's Shocked and Amazed! On and Off the Midway,* vol. 5, 65–69. Baltimore: Dolphin-Moon Press/Atomic Books.

Reese, Ralph. 1996. The art of gaffing freaks. In *The Big Book of Freaks,* edited by Gahan Wilson et al., 189–91. New York: Paradox Press.

"Science of Magic." 1997. Documentary on Discovery Channel, aired 30 November.

Taylor, James. 1997. *James Taylor's Shocked and Amazed! On and Off the Midway,* vol. 4. Baltimore: Dolphin-Moon Press/Atomic Books.

———. 1998. *James Taylor's Shocked and Amazed! On and Off the Midway,* vol. 5. Baltimore: Dolphin-Moon Press/Atomic Books.

Teller [of Penn and Teller, magicians]. 1997. Gorilla girl. In *James Taylor's Shocked and Amazed! On and Off the Midway,* vol. 4. Baltimore: Dolphin-Moon Press/Atomic Books.

12

"Mothman" Solved!

Investigating on Site

The ill-fated 2002 movie *The Mothman Prophecies*, based on a book of the same title (Keel 1975), focused on a "flying monster" that plagued the Point Pleasant, West Virginia, area for a year, beginning in November 1966. In addition to giant-bird sightings, the tale involved alien contacts, Men in Black, a tragic bridge collapse, and other elements (Nickell 2002). In April 2002 I was able to make an investigative trip to Point Pleasant, spending a few days there. I came back with some interesting and illuminating information on the case.

Pranklore

A popular legend of the Point Pleasant area holds that "Mothman" was the creation of a prankster. Supposedly, a local man dressed in a Halloween costume had hidden at the abandoned munitions complex known as the TNT area, about five miles north of Point Pleasant, and had scared young couples by jumping out at their cars at night.

But this local legend is not credible, in my opinion. For one thing, knowledgeable area residents call attention to the fact that there have been several different claimants of the prankster title. For example, Rush Finley (2002), who with his wife, Ruth, owns the historic Lowe Hotel where I stayed, told me there were "at least half a dozen people" who now claim responsibility for the pranks, supposedly done when they were teenagers. Finley is echoed by Charlie Cline (2002), manager of the music store Criminal Records, who thinks many are "jumping on the bandwagon" in this regard. Cline has heard several such stories.

Another reason this explanation is not credible (except perhaps for later, bandwagon or copycat pranks) is that the appearance of such a trickster is not at all compatible with the original eyewitnesses' descriptions of the creature, especially with regard to its glowing red eyes (as we shall see presently).

Costumed prankster or not, there *were* Mothman hoaxes. Rush Finley told me how some construction workers had used helium from welding tanks to make balloons from sheet plastic and tied red flashlights to them one night. Thus weighted, these Mothmen did not soar high but only drifted over the treetops.

Still other pranks and hoaxes occurred following the first wave of sightings. The spring of 1967 brought a number of UFO reports that were described in local newspapers and involved both misidentification of mundane phenomena and deliberate hoaxing. Some of the UFOs were soon identified as commercial or military planes (notably a U.S. C-119 "flying boxcar" on a training mission from Columbus, Ohio). However, a private plane with a "prankster pilot" was reported to have been "gliding back and forth across the river for several nights" to frighten locals. On one occasion, however, according to a newspaper account, the pilot came too close to a hilltop and was suddenly forced "to cut his engines on." (See newspaper clippings in Sergent and Wamsley 2002.)

The Sightings

Detailed reports demonstrate that the original Mothman sightings were not hoaxes. They occurred on Tuesday night, November 15, 1966, when two couples drove through the TNT area and saw a winged creature, "shaped like a man, but bigger," with glowing red eyes. It walked on sturdy legs with a shuffling gait and, when it took flight and seemed to follow them, it "wasn't even flapping its wings." They said it "squeaked like a big mouse" (quoted in Keel 1975, 59–60).

Other reports soon flooded in, including one from two Point Pleasant firemen who visited the TNT area only three nights after the first sighting. They saw the "huge" creature but were positive that "[i]t was definitely a bird." Many witnesses described it as headless, with shining red eyes set near the top of its body, although one woman spoke of its "funny little face" (Keel 1975, 64–65, 71).

Despite accounts of "glowing" eyes, one of the original eyewitnesses, Linda Scarberry, specifically stated that the effect was related to the car headlights. "There was no glowing about it until the lights hit it," she said. (This statement is part of her handwritten account of the incident from 1966, reproduced in Sergent and Wamsley 2002, 36–59). Others echoed her statement. For example, one man, alerted by his dog, aimed his flashlight in the direction of his barn, "and it picked up two red circles, or eyes, which," he said, "looked like bicycle reflectors" (Keel 1975, 56).

Eyeshine

The reflector-like nature of the creature's eyes is revealing. As ornithologists well know, some birds' eyes shine bright red at night when caught in a beam from auto headlights or a flashlight. "This 'eyeshine' is not the iris color," explains an authority, "but that of the vascular membrane—the tapetum—showing through the translucent pigment layer on the surface of the retina" (Gill 1994, 188).

Now, the TNT area, which I visited during both days and nights, is surrounded by the McClintic Wildlife Management Area—then as now a bird sanctuary! Owls, which exhibit crimson eyeshine, populate the area. Indeed, Steve Warner (2002), who works for West Virginia Munitions to produce .50-caliber ammunition in the TNT compound, told me there were "owls all over this place." Interestingly, neither he nor a coworker, Duane Chatworthy (2002), had ever seen Mothman, although Warner pointed out he had lived in the region all his life.

Because of Mothman's squeaky cry, "funny little face," and other features, including its presence near barns and abandoned buildings, I identified it as the common barn owl (Nickell 2002). (See Figure 12-1.) One reader (Long 2002) insisted it was instead a great horned owl which, although it does not match certain features as well, does have the advantage of larger size. It seems likely that various owls and even other large birds have played Mothman on occasion.

I did some further research regarding eyeshine, learning that the barn owl's was "weak" and the great horned owl's only "medium." However, the barred owl exhibits "strong" eyeshine (Walker 1974, 218–22) and—according to David McClung (2002), wildlife manager at McClintic—is common to the area; indeed, it is even more prevalent there than the barn owl. It is also larger than the barn owl, which it somewhat resembles, and is "only a little smaller than the Great Horned Owl" (Kaufman 1996, 317). (Mounted specimens of these and other owl species are profusely displayed in the West Virginia State Farm Museum near the McClintic preserve. Museum director Lloyd Akers generously allowed me special access to examine and photograph them.)

In light of the evidence, it seems very likely that the Mothman sightings were mostly caused by owls—probably more than one type. A man named Asa Henry shot and killed one, tentatively identified as a snowy owl, during the Mothman flap. Although only about two feet tall, a newspaper dubbed it a "giant owl" due to its wingspan of nearly five feet (Sergent and Wamsley 2002, 94, 99). In Point Pleasant I was able to

FIGURE 12-1. Split-image drawing compares Mothman (left) to common barn owl (illustration by Joe Nickell).

view the mounted specimen and to speak with Mr. Henry's grandson, David Pyles. Himself a taxidermist, Pyles (2002), who is "very skeptical of Mothman," told me his grandfather had always maintained that the Mothman furor ended after he shot the bird.

"Bighoot"

Owls are very likely responsible for other birdman sightings. One of the these is the 1952 case of the Flatwoods Monster that supposedly arrived in Flatwoods, West Virginia, aboard a flying saucer. Loren Coleman, in his *Mothman and Other Curious Encounters* (2002), sees in that case "elements foreshadowing" the subsequent Mothman reports. However, as a Michigan Audubon Society publication concluded, my investigative report on the case (Nickell 2000) "convincingly demonstrates that the alleged flying saucer was really a meteor and the hissing creature from outer space was none other than a Barn Owl! Check it out, it's a real scream!" ("Those Monster Owls" 2001).

Somewhat similarly, several 1976 sightings in Cornwall, England, featured a "big feathered bird man" that was first seen "hovering over a church tower"—a common nesting place for barn owls (Kaufman 1996, 306). Appropriately, the entity became known as "Owlman" (Coleman 2002, 34–36).

As to Mothman, "cryptozoologist" Mark A. Hall (1998) has opined that it may be a hitherto undiscovered species of giant owl. He has dubbed it "Bighoot" and cites evidence that it has long existed in the Point Pleasant area. I take this as an implicit concession that Mothman—of all the creatures known to science—most resembles an owl, except for size.

Here, then, is the question that separates the mystifiers from the skeptics: Is it more likely that there has long been a previously undiscovered giant species among the order *strigiformes* (owls), or that some people who suddenly encountered a "monster" at night misjudged its size? The latter possibility is supported by the principle of Occam's razor, which holds that the simplest tenable explanation is to be preferred as most likely correct. The principle seems especially applicable to the case of Mothman.

REFERENCES

Chatworthy, Duane. 2002. Interview by author, 12 April.
Cline, Charlie. 2002. Interview by author, 12 April.
Coleman, Loren. 2002. *Mothman and Other Curious Encounters*. New York: Paraview Press.
Finley, Rush. 2002. Interview by author, 12 April.
Gill, Frank B. 1994. *Ornithology*. 2d ed. New York: W. H. Freeman.
Hall, Mark A. 1998. Bighoot—the giant owl. *Wonders* 5, no. 3 (September): 67–79 (cited in Coleman 2002).
Kaufman, Kenn. 1996. *Lives of North American Birds*. New York: Houghton Mifflin.
Keel, John A. [1975] 1991. *The Mothman Prophecies*. Reprinted New York: Tor.
Long, Chris. 2002. Letter to editor. *Skeptical Inquirer* 26, no. 4 (July/August): 66.
McClung, David. 2002. Interview by author, 13 April.
Nickell, Joe. 2000. The Flatwoods UFO monster. *Skeptical Inquirer* 24, no. 6 (November/December): 15–19.
———. 2002. "Mothman" solved! *Skeptical Inquirer* 26, no. 2 (March/April): 20–21.
Pyles, David. 2002. Interview by author, 12 April.
Sergent, Dorrie, Jr., and Jeff Wamsley. 2002. *Mothman: The Facts Behind the Legend*. Point Pleasant, W. Va.: Mothman Lives Publishing.
Those monster owls. 2001. *The Jack-Pine Warbler* (March/April): 6.
Walker, Lewis Wayne. 1974. *The Book of Owls*. New York: Alfred A. Knopf.
Warner, Steve. 2002. Interview by author, 12 April.

13

Relics of the Headless Saint

For centuries, the site of the Cathedral of Santiago de Compostela in northwestern Spain has been a place of reputed miracles, including revelations and healings. Today, among its visitors are New Agers who consider the cathedral "a reservoir of powerful positive psychic energy"; some even claim to see apparitions of earlier pilgrims (Hauck 2000, 133–34).

On September 6, 1997, I made my own "pilgrimage" to the historic cathedral. I had been attending the Ninth EuroSkeptics Conference in the nearby seaport city of La Coruña ("The Crown"), and the cathedral was the focus of one Saturday's scheduled sightseeing trip—a secular pilgrimage in the company of scientists and other skeptics, including CSICOP's chairman Paul Kurtz and executive director Barry Karr. Not only did I appreciate the cathedral's Romanesque architecture, but I also began to delve into its history, steeped in centuries-old myths and pious legends.

Legends of St. James

The cathedral marks the site of the allegedly miraculous discovery of the remains of St. James the Greater, so named to distinguish him from the other apostle of that name. (There were various other Jameses in the Christian gospels as well, including one of Jesus's brothers [Mark 6:3; Matthew 13:55]). James the Greater was a son of Zebedee. Jesus found this James and his brother John mending nets by Lake Genesareth (also known as the "Sea of Galilee") and called them to his ministry. (This was just after he had similarly invited Simon and Andrew, promising to make them "fishers of men" [Mark 1:16–20].) In the early history of the church, James was the first disciple to be martyred (Acts 12:1–2): he was executed by King Herod Agrippa I in 44 C.E. According to one legend, his accuser repented as the execution was about to occur and was beheaded along with James (Jones 1994).

By the seventh century, another pious legend claimed that James had taken the gospel to Spain. Subsequently, still another legend told how Herod had forbidden the burial of James's beheaded body. Therefore, on the night after the execution, several Christians secretly carried his remains to a ship. "Angels" then conducted the vessel "miraculously" to Spain, and the body was transported to the site of the present-day cathedral.

The apostle's body lay undiscovered until the early ninth century (about 813 C.E.). Then, according to still another miracle tale, a pious friar was led to the site by a "star," in much the same manner as the Wise Men were supposedly guided to the birthplace of Jesus in the New Testament (Matthew 2:1–12). The supernatural light revealed the burial place. The local bishop accepted the validity of the friar's discovery and had a small basilica built over the supposed saint's sepulcher. It was destroyed a century later in a Muslim raid, but in 1078 work was begun on the present cathedral, and was largely completed in 1128 (McBirnie 1973; Coulson 1958; *El Camino* 1990, 2–3, 36–37).

The alleged discovery came at an opportune time. After the Moors conquered Spain, only its northwest corner remained independent, and it was from there that the drive to reconquer the country for Christendom was launched (*Encyclopedia Britannica* 1960). The supposedly divine revelation of the relics seemed to endorse the quest, and Saint James (Santiago) "became the rallying figure for Christian opposition to the Moors" (Jones 1994, 144). Miracles began to occur at the site, resulting in "an extensive collection of stories" that were "designed to give courage to the warriors" fighting against the Moors. There were even stories of the saint appearing on the battlefield at crucial moments, and he was sometimes known as *Santiago Matamoros* ("St. James the Moorslayer"). The tales also encouraged the pilgrims who were beginning to wend their way to Compostela (*El Camino* 1990, 3, 36–37; Cavendish 1989, 251).

Magical Relics

Medieval pilgrims were attracted to holy places, including churches that were home to powerful relics. In Catholicism, a *relic* is an object associated with a saint or martyr (a bone, piece of clothing, etc.). According to Kenneth L. Woodward's *Making Saints* (1990, 63): "Just as the soul was totally present in every part of the body, so, it was popularly believed, the spirit of the saint was powerfully present in each relic. Thus, detached from the whole body and separated from the tomb, relics took on magical power of their own." As well, mere proximity could be enough: "[T]he medieval pilgrim was satisfied if he could but gaze on the tomb of his cult-object" (Pick 1929, 101–2).

At the Cathedral of Santiago de Compostela, the relics of St. James were reputed to be working wonders, and various "prodigies, miracles and visions" multiplied there (*El Camino* 1990). Many made the pilgrimage to Santiago to be healed of an affliction (Gitlitz and Davidson 2000), but perhaps most did so for the experience itself and to be compensated with indulgences (remissions of punishment due a sinner) (*El Camino* 1990, 4). "In the Middle Ages," notes one reference work (Kennedy 1984, 93), "it was possible for the faithful to buy such pardon for their sins, and unscrupulous priests saw the selling of indulgences as an easy way of raising money." (Abuse of indulgences was among the criticisms that led to the Protestant Reformation.)

Pilgrimages

There were many roads to Compostela, but one of three major medieval pilgrimage routes remains popular today, inviting not only religious supplicants but also historians, art lovers, adventurers, and others. It is a nearly 500-mile trek called the Santiago de Compostela Camino—or just *Camino* for short (meaning "the road" or "way"). Beginning in France, it winds across the Pyrenees, then traverses northern Spain westward to Santiago (Gitlitz and Davidson 2000; MacLaine 2000).

Among those who made pilgrimages to Compostela were such historic notables as St. Francis of Assisi and the Spanish monarchs Ferdinand and Isabella. More recently, there was Shirley MacLaine. The acclaimed actress—sometimes disparaged as "the archetypal New Age nut case" for her belief in past lives, alternative medicine, and other fringe topics (Neville 2000)—wrote about her experiences along the pilgrims' way in *The Camino: A Journey of the Spirit*. Although the book was published in 2000, she actually made the trek in 1994, as shown by her *compostelana* (pilgrim certificate of Santiago de Compostela) reproduced inside the book's covers.

Camino is not merely a record of MacLaine's journeys (both inward and outward) but a veritable catalog of her mystical notions and fantasy experiences. Never religious, she says, she adopted a New Age mantra of "opting instead to seek spirituality." On her trip she soon felt that she was being "visited by an angel named Ariel," which, she wrote, "began to talk to me in my head," telling her to "'Learn to have pleasure as you experience it.'" In the book she reflects on some of her alleged past-life memories or "revisitations": as a geisha in Japan and as a dimly recalled inhabitant of India and again Russia. Later, when she is "not really dreaming," she has an "intensely real" experience in which a monk appears to her, announcing, "I am John the Scot." She is able to converse with him about the "science" of astrology, and about "karma" (supposedly the consequences of a person's deeds that carry over and help shape his or her next reincarnation). They also discuss "ley lines" (imaginary connections between supposed sites of power, such as megaliths, ancient monuments, holy wells, temples, etc.) (Guiley 1991).

Occasionally reality intrudes. Once, seeking to relieve herself, MacLaine noted that she squatted over an anthill! Sometimes she is admirably observant:

> In every village I was awed by the opulent richness of the churches, while the poor people who attended them gave every last penny they had to the collection plate. One priest sold holy candles to the peasants, which they lit, placed on the altar, and prayed over. When they left, the priest put them up for sale again. They had paid for the privilege of praying.

Midway through her narrative, MacLaine pauses to decide "whether to include the ensuing events" that promise to take the reader "off the Camino path and to the edge of reason." Then she launches into another of her "dream-visions" in which John the Scot guides her on an odyssey to the legendary lost continents of Lemuria and Atlantis. There she learns that the latter was "an advanced colony of Lemurians" and that they in turn had received input from extraterrestrials who have been surveying Earth for millennia (MacLaine 2000, 187, 213).

Eventually, she arrives at the cathedral to pay her respects to Santiago de Compostela—Saint James—or as she describes him, "the saint with no head," adding, "I felt the same way" (MacLaine 2000, 294). There, in a practice familiar to countless visitors, she climbed the stairs leading behind a Romanesque painted-stone statue of the seated apostle and, as directed by custom since the seventeenth century (Gitlitz and Davidson 2000, 344), hugged the effigy (see Figures 13-1 to 13-3). She does not mention whether she then descended to the crypt to view the reliquary that supposedly contains the saint's bones, but—after a priest ritualistically bathed her feet—she was soon out of the cathedral and on a flight to Madrid. She reflects on her odyssey of imagination and reality: "Perhaps all of it is simple. We came from the Divine; we create with that imaginative energy until we return to it. Lifetime after lifetime" (MacLaine 2000, 306).

Although the Camino pilgrimage has declined over the past few centuries, one reviewer predicted that MacLaine's book "will change that" (Neville 2000). Even before MacLaine's visit, though, the number of pilgrims had begun to multiply once again. Now, according to *The International Directory of Haunted Places* (Hauck 2000, 133–34), New-Age visitors to the cathedral consider it "a reservoir of powerful positive psychic energy." Indeed, "[t]hey sense the spiritual energy and devotion to the divine that tens of thousands of pilgrims brought to this site, and sometimes they even report seeing the apparitions of those dedicated souls making their way through the city to the holy shrine."

Investigation

The relics of St. James are the central focus of the shrine, and indeed its very *raison d'être*—but are they genuine? If they are not, what does that say about all the reputed miracles there: the alleged revelations, healings, apparitions, and other supernatural and paranormal phenomena?

Today few historians believe that St. James ever visited Spain. According to one guidebook, *Roads to Santiago* (Nooteboom 1997, 201): "The fiery resplendence of Santiago and all it inspired came about be-

FIGURES 13-1, 13-2, 13-3.

In Spain's Cathedral of Santiago de Compostela, pilgrims climb stairs to an alcove behind the statue of Santiago (St. James), which they embrace in a centuries-old tradition. (Photographs by Joe Nickell.)

cause people *believed* they had found the grave of the apostle James in that town, events therefore that were set in motion by something that perhaps never took place at all." One dictionary of saints (Coulson 1958, 237) explains some of the reasons for skepticism:

> Tradition asserts that James brought the gospel to Spain, but because of the early date of his death, this claim is quite untenable. In the Acts of the Apostles it is Paul who is depicted as the pioneer missionary, and James was dead before Paul's activity began.
>
> In fact, the tradition only appears in written form for the first time in the seventh century, arising from a Greek source of doubtful historical credentials, but it was a century later, when a star miraculously revealed what was claimed to be the tomb of James, that popular belief spread. This shrine at Compostella (probably derived from Campus stellae: the field of the star) rivalled Rome as a center of pilgrimage.

But did the legend of the star give rise to the name *Compostela*, or was it the other way around? According to an official Spanish government guidebook (*El Camino* 1990, 2), the place chosen for deposit of St. James' sarcophagus was "at exactly the spot where a former *compostum*—cemetery—lay which in the course of time became Compostela."[1] In fact, excavations beneath the cathedral have yielded "remnants of a pre-Roman necropolis" as well as "[r]emains of a Roman cemetery," together with "an altar dedicated to Jupiter" (Gitlitz and Davidson 2000, 346, 351). In light of these facts, it seems plausible that the name *Compostela* might have derived not from *Campus stellae* ("star field") but from *Campus stelae* (a *stele* being an inscribed stone), that is, "field of monuments" or "gravestone field." Another possibility is that the name is a combined form of *compositus* ("orderly arrangement") and *stelae* ("tombstones").

Even more likely (according to Kevin Christopher, CSICOP's publicity director, who has degrees in classics and linguistics) is the possibility that *compostela* is simply a diminutive form of *compostum*.

If any of these alternate interpretations is correct, it would suggest that a name that originally meant "graveyard" was mistranslated as "star field," and that the mistranslation in turn prompted the little tale purporting to "explain" the name. The process by which a folk etymology apparently leads to the creation of a legend is well known. One example is the name of a British tribe, Trinovantes, which seems to have been falsely attributed to Troynovant, or "New Troy," and so to have prompted a legend that remnant Trojans settled a then-uninhabited Britain (Howatson 1989, 582–83). Another case in point is the name of a class of "miraculous" Christ portraits, *vera icona*, or "true images," that became known as "veronicas"—hence apparently inspiring the legend that a pious woman of that name gave her veil to Jesus to wipe his face as he struggled to his crucifixion (Nickell 1998, 19–29, 73–77).

As we have seen, there were numerous other legends about St. James, as of course there were about other religious figures and subjects. Many factors contributed to the manufacture of saints' legends. For example, speaking specifically of Santiago, one source observes that "many of the great romances of the middle ages developed from the tales told by the pilgrims to while away the tedium of the long journey to this remote corner of Spain" (*Encyclopedia Britannica* 1960). A more sinister view of the entire affair regarding St. James and his legends is given by Hauck (2000, 133–34):

> The discovery of his relics was apparently a hoax perpetrated by the Church to attract pilgrims and take the region back from Arabian settlers [the Moors]. It is known that the Cathedral of Santiago sent hired "storytellers" to spread the news of miracles associated with the relics, and their tactics seem to have worked, for by the

twelfth century, this was the most popular pilgrimage site in Europe.

Of course, discrediting the legend of the relics' miraculous discovery, and even debunking the alleged missionary work of St. James in Spain, "does not dispose of the claim that the relics at Compostela are his" (Coulson 1958, 237). Yet how likely is it that the apostle's remains would have been arduously transported to northern Spain in the first place, and then have remained unknown until they were allegedly revealed nearly eight centuries later?

Additional doubts are raised by the fact that the remains of St. James, when discovered, were accompanied by the skeletons of two others. Though that would not be surprising at the site of an ancient cemetery, how did the pious legendmakers explain the two extra bodies buried with the apostle? They simply declared them to be the relics of "two of his disciples" (McBirnie 1973, 94).

Further suspicion about the authenticity of the relics comes from the climate of relic-mongering that was prevalent in the Middle Ages. As the demand for relics intensified, "a wholesale business in fakes" grew in response (Pick 1929, 101–2). Alleged relics included the fingers of St. Paul, John the Baptist, and the doubting Thomas. Most prolific were "relics" associated with Jesus himself. No fewer than six churches preserved his foreskin. There were also bits of hay from the manger in which he was laid at birth, gifts from the Wise Men, and vials of Mary's breast milk. From the crucifixion, various churches had thorns from the crown of thorns, although the Sainte Chapelle in Paris possessed the entire object. There were more than 40 "true" shrouds, including the notorious Shroud of Turin, which appeared in the middle of the fourteenth century as part of a faith-healing scheme. (See chapter 22, "Scandals and Follies of the 'Holy Shroud.'")

In the case of St. James, there is even a question about the exact nature of the relics. McBirnie (1973, 106–7), in his *Search for the Twelve*

Apostles, declares it a certainty that James's body was buried in Jerusalem. However, he believes it possible that later "some of the bones of the Apostle, perhaps the body" might have been removed to Spain, with the head left behind in Jerusalem. Contravening stories surround alleged portions of James's body that are housed elsewhere. For example, at Constantinople a shrine held "a silver arm encompassing a relic of St. James the Greater," which was taken to Troyes, France after the capture of Constantinople in 1204 (Gies and Gies 1969, 128). Another relic, the saint's hand, is supposedly preserved at the abbey in Reading, England (Jones 1994). Still another relic is claimed by an Italian cathedral (McBirnie 1973, 96).

Of course, the relics could have been subdivided, following a common medieval practice (McBirnie 1973, 107), but even the presumed link between the relics that were supposedly revealed miraculously in the early ninth century, and those enshrined at Compostela today, is questionable. As Gitlitz and Davidson (2000) report:

> Actually, Santiago's bones were hidden several times in successive centuries to keep them out of the hands of various threatening parties, such as Drake, who wanted them for England, and various Spanish monarchs, who coveted them for the Escorial. Eventually their exact location was forgotten altogether, although pilgrims continued to venerate an urn on the altar that they believed held the bones. Excavations in 1878–9 unearthed some bones that—when the discoverer went temporarily blind—were held to be those of the Apostle. Six years later Pope Leo XIII issued a bull verifying the validity of the relics, thus—at least officially—ending all controversy.

In short, there are some bones at Compostela the provenance of which cannot credibly be traced to James the Greater. As *The Penguin*

Dictionary of Saints concludes, there is "no evidence whatever as to the identity of the relics discovered in Galicia early in the ninth century and claimed to be those of St. James" (Attwater 1983, 179). But if the relics are bogus, as the evidence strongly indicates, how can we explain the reported supernatural and paranormal events there? Can they actually have naturalistic explanations? Indeed they can. For example, supposedly divine cures may simply be due to the body's own healing ability eventually proving effective, or to an abatement of the illness (known as spontaneous remission), or to the effects of suggestion (the well-known placebo effect). A reduction in pain, whether caused by suggestion or by the physiological effects of excitement, may also give the illusion that a miracle "cure" has taken place (Nickell 1998).

Visionary experiences, like those of Shirley MacLaine, can also have prosaic explanations. They may be due to pilgrims' heightened expectations and to many other factors, including the propensity of certain individuals to fantasize. MacLaine herself exhibits several traits associated with the "fantasy-prone" personality: Such persons often have rich fantasy lives, believe they have psychic powers, supposedly receive special messages from higher beings, and report vivid dreams and apparitional experiences (Wilson and Barber 1983; Baker and Nickell 1992, 221–26).

The "dream-visions" that MacLaine had when she rested along the Camino—experiences that "seemed more than a dream" (MacLaine 2000, 79, 105)—may have been what is termed *lucid dreaming*. A lucid dream is one in which the dreamer is able to direct the course of the dream, "something like waking up in your dreams" (Blackmore 1991). In fact, MacLaine (2000, 59) says she "realized that on some level I must be controlling in some manner what I dreamed." That she had her dream-visions while hiking nearly 500 miles at a rate of up to 20 miles a day is interesting, as it is known that lucid dreaming tends to occur following "high levels of physical (and emotional) activity" (Blackmore 1991, 365).

Similar explanations may apply to the apparitions reported by some New-Age visitors to Santiago de Compostela: ghostly pilgrims seen "making their way through the city to the holy shrine" (Hauck 2000, 133–34). Such apparitions may be nothing more than mental images superimposed on the actual visual scene—especially if the one seeing them is daydreaming or performing some routine activity (Nickell 2000), conditions consistent with walking on a long journey.

As to any protective powers supposedly obtained by hugging the saint's statue, my own experience belies any such notion. After performing the charming ritual myself, I stepped up onto a narrow ledge of the small chamber to attempt a better view for some snapshots. Although I survived that precarious act, later in the afternoon I slipped on steps outside my hotel in La Coruña and broke my leg very badly. (Picture my lying in agony surrounded by skeptics, who suggest I may only have a sprain and invite me to see if I can wiggle my toes. But when I lift my leg and they observe the bizarre angle of my foot, their doubts are silenced!)

Whether or not the Cathedral of Santiago de Compostela houses a saint's relics, which may or may not exude supernatural power, the cathedral is nevertheless a monument to the persistence of magical thinking. Apparently built on the site of a Roman shrine to Jupiter (with accompanying necropolis), it became a focal point for Christian pilgrims, and now seems to be undergoing further transition as New Agers adapt its legends and history to their own occultish superstitions. They see it as a site of powerful "psychic energy" and the pilgrimage route as a tracing of mystical "ley lines" (Hauck 2000, 133–34; MacLaine 2000, 4–5). Although specific beliefs change, what Paul Kurtz terms "the quest for transcendence" seems perpetual (1991, 23–26).

NOTE

1. The usual Latin word for cemetery is *sepulcretum* (or *sepulcrum*, a place of interment). However, *compono* (with forms *compostus*, *compositum*, etc.) means "to lay out for burial, place in an urn, bury." See *Oxford Latin Dictionary* 1969.

REFERENCES

Attwater, Donald. 1983. *The Penguin Dictionary of Saints*. London: Penguin.

Baker, Robert A., and Joe Nickell. 1992. *Missing Pieces: How to Investigate Ghosts, UFOs, Psychics, & Other Mysteries*. Buffalo, N.Y.: Prometheus Books.

Blackmore, Susan. 1991. Lucid dreaming: Awake in your sleep? *Skeptical Inquirer* 15, no. 4 (Summer): 362–70.

Cavendish, Richard, ed. 1989. *Legends of the World*. New York: Crescent Books.

Coulson, John, ed. 1958. *The Saints: A Concise Biographical Dictionary*. New York: Hawthorne Books.

El Camino de Santiago. 1990. Travel booklet. N.p., Spain: Ministerio de Transportes, Turismo y Comunicaciones.

Encyclopedia Britannica. 1960. s.v. "Santiago de Compostela."

Gies, Joseph, and Frances Gies. 1969. *Life in a Medieval City*. New York: Harper Colophon Books.

Gitlitz, David M., and Linda Kay Davidson. 2000. *The Pilgrimage Road to Santiago*. New York: St. Martin's Griffin.

Guiley, Rosemary Ellen. 1991. *Harper's Encyclopedia of Mystical & Paranormal Experience*. s.v. "Karma," "Leys." New York: HarperCollins.

Hauck, Dennis William. 2000. *The International Directory of Haunted Places*. New York: Penguin Books.

Howatson, M. C., ed. 1989. *The Oxford Companion to Classical Literature*. 2d ed. Oxford: Oxford University Press.

Jones, Alison. 1994. *The Wordsworth Dictionary of Saints*. Hertfordshire, England: Wordsworth Reference.

Kennedy, Richard. 1984. *The Dictionary of Beliefs*. London: Ward Lock Educational.

Kurtz, Paul. 1991. *The Transcendental Temptation: A Critique of Religion and the Paranormal*. Buffalo, N.Y.: Prometheus Books.

MacLaine, Shirley. 2000. *The Camino: A Journey of the Spirit*. New York: Pocket Books.

McBirnie, William Steuart. 1973. *The Search for the Twelve Apostles*. Wheaton, Ill.: Tyndale House, 87–107.

Neville, Anne. 2000. Walking a mile in the pilgrim's shoes. Review of *The Camino: A Journey of the Spirit*, by Shirley MacLaine. *Buffalo News* (Buffalo, N.Y.), 5 July.

Nickell, Joe. 1998. *Looking for a Miracle*. Amherst, N.Y.: Prometheus Books.

———. 2000. Haunted inns: Tales of spectral guests. *Skeptical Inquirer* 24, no. 5 (September/October): 17–21.

Nooteboom, Gees. 1997. *Roads to Santiago: A Modern-Day Pilgrimage through Spain*. New York: Harcourt.

Oxford Latin Dictionary. 1969. s.v. "compono."

Pick, Christopher, ed. 1929. *Mysteries of the World.* Secaucus, N.J.: Chartwell Books.

Wilson, Shirley C., and T. X. Barber. 1983. The fantasy-prone personality. In *Imagery: Current Theory, Research and Application*, edited by A. A. Sheikh. New York: John Wiley & Sons.

Woodward, Kenneth L. 1990. *Making Saints*. New York: Simon & Schuster.

14

Circular Reasoning

Crop Circles and Their "Orbs" of Light

Since they began to capture media attention in the mid-1970s, and throughout their proliferation and evolution during the decades of the 1980s and 1990s, crop circles have generated mystery and controversy. New books touting "scientific research" continue the trend. The topic also got a boost from a 2002 Hollywood movie, *Signs*, starring Mel Gibson as a Pennsylvania farmer who discovers a 500-foot design imprinted in his crops and seeks to learn its meaning.

At issue are swirled, often circular designs pressed into crop fields, especially those of southern England. They can range from small circles only a few feet in diameter to elaborate "pictograms," some now as large as a few hundred feet across. By the end of the 1980s, books on the crop-circle phenomenon had begun to spring up as well, and soon circles-mystery enthusiasts were being dubbed *cereologists* (after Ceres, the Roman goddess of vegetation). Circlemania was by then in full bloom (Delgado and Andrews 1989; Nickell and Fischer 1992).

Most cereologists—also known as "croppies" (Hoggart and Hutchinson 1995)—believed the circular designs were being produced either by extraterrestrials or by hypothesized "plasma vortices," which are supposedly "small, local whirlwinds of ionized air" (Haselhoff 2001, 5–6). A few took a more mystical approach. When I visited the vast wheat crops of the picturesque Wiltshire countryside in 1994, at one formation a local dowser told me he believed the swirled patterns were produced by spirits of the earth (Nickell 1995).

Hoaxers, most croppies insisted, could not be responsible because the plants were only bent and not broken, and there were no footprints or other traces of human activity. Skeptics replied that from mid-May to early August, English wheat is green and pliable, and can be broken only with difficulty. As for tracks, they were precluded by *de facto* footpaths in the form of the tractor "tramlines" that mark the fields in closely spaced, parallel rows (Nickell and Fischer 1992).

Investigation into the circles mystery indicated that it might be profitable to look not just at individual formations but rather at the overall phenomenon (with a nod to the old principle that one may fail to see the forest for the trees). Forensic analyst John F. Fischer and I soon identified several characteristics that suggested the work of hoaxers (Nickell and Fischer 1992):

1. *An escalation in frequency.* Although there had been sporadic reports of simple circles in earlier times and in various countries (possibly as UFO-landing-spot hoaxes), the "classic" crop circles began to be reported by the mid-1970s. Data on the circle reports showed that their number increased annually from 1981 to 1987, an escalation that seemed to correlate with media coverage of the phenomenon. In fact, it appeared that the coverage helped prompt further hoaxes.

2. *Geographic distribution.* The phenomenon showed a decided predilection for a limited geographic area, flourishing in southern England: Hampshire, Wiltshire, and nearby counties. It was there that the circles effect first captured the world's attention, but just as the number of circles increased, so their locations also spread. After newspaper and television reports on the phenomenon began to proliferate in the later 1980s, the formations began to crop up (so to speak) in significant numbers around the world. Indeed, the circles effect appeared to be a media-borne "virus."
3. *Increase in complexity.* A very important characteristic of the patterned-crops phenomenon was the tendency of the configurations to become increasingly elaborate over time. They progressed from simple swirled circles to circles with rings and satellites, to still more complex patterns. In 1987 came a crop message, "WEARENOTALONE" (although skeptics observed that, if the source were indeed English-speaking extraterrestrials, the message should have read "You" rather than "We"). In 1990 came still more complex patterns, dubbed "pictograms." There were also free-form shapes, such as a tadpole-like design, a witty crop *triangle*, and the hilarious *bicycle* (see Hoggart and Hutchinson 1995, 59).

   There also appeared beautifully interlinked spirals, a Menorah, intricate snowflake and stylized spider-web designs, elaborate torus-knot and mandala emblems, pentagram and floral patterns, and other distinctive formations, including an "origami hexagram" and several fractals (mathematical designs with a motif subjected to repeated subdivision)—all consistent with the intelligence of modern *homo sapiens*. At the end of the decade came many designs that included decidedly square and rectilinear shapes, seeming to represent a wry response to the hypothesized swirling "vortex" mechanism.
4. *The shyness factor.* A fourth characteristic of the crop-field phenomenon is its avoidance of observation. It is largely nocturnal,

> and the designs even appear to specifically resist being seen, as shown by Operation White Crow. That was an eight-night vigil maintained by about 60 cereologists in June 1989. Not only did no circles appear in the field chosen for surveillance but—although almost 100 formations had already appeared that summer, with yet another 170 or so to occur—*not a single circle was reported during the observation period anywhere in England.* Then a large circle-and-ring formation was discovered about 500 yards away *on the very next day!*

These and other characteristics are entirely consistent with the work of hoaxers. Indeed, as John Fischer and I were about to go to press with our investigative report, in September 1991, two "jovial con men in their sixties" confessed that they had been responsible for many of the crop formations made over the years. In support of their claim, the men, Doug Bower and Dave Chorley, fooled cereologist Pat Delgado. He declared a pattern they had produced for a tabloid to be authentic, insisting that it was of a type no hoaxer could have made. The pair used a rope-and-plank device to flatten the plants, and even demonstrated their technique for television crews; one such demonstration was aired on ABC-TV's *Good Morning America* on September 10, 1991 (Nickell and Fischer 1992, 145–48). (See Figures 14-1 and 14-2.)

Cereologists were forced to concede that hoaxers were producing elaborate designs and that "there are many ways to make a hoaxed crop circle" (Haselhoff 2001, 34). (For example, some who go round in circles use a garden roller to flatten the plants [Hoggart and Hutchinson 1995].) Whereas in the past some cereologists thought they could distinguish "real" from fake circles by dowsing (Nickell 1995), the more cautious now admit it is not an easy matter, "certainly not as long as we do not even know exactly what mechanism creates crop circles" (Haselhoff 2001, 34).

Nevertheless, the croppies were sure that some of the formations

FIGURE 14-1. Crop-circle design in a field of oats in upstate New York.

FIGURE 14-2. Circlemakers who produced the crop circle in Figure 14-1: Kevin Christopher, Benjamin Radford, and Joe Nickell. (Phots by Benjamin Radford.)

must be genuine, citing various "unexplained" features. More recently they have invoked new "scientific" evidence in that regard, such as that provided by "the BLT Research Team" in Cambridge, Massachusetts. The "B" and "T" are circle "researchers" and "L" is a semi-retired biophysicist, W. C. Levengood. He finds a correlation between certain deformities in plants and their locations within crop-circle-type formations, but not in control plants outside them (Levengood and Talbott 1999). However, correlation is not causation, and there are other objections to his work (Nickell 1996a). As well, more mundane hypotheses for the effects—for instance, compressed *moist* plants steaming in the hot sun—appear to have been insufficiently considered.

Crucially, because there is no satisfactory evidence that a single "genuine" (i.e., "vortex"-produced) crop circle exists, Levengood's reasoning is circular: although there are no guaranteed genuine formations on which to conduct research, the research supposedly proves the genuineness of the formations. Furthermore, if his work were really valid, Levengood should have found that a high percentage of the crop circles chosen for research were actually hoaxed, especially since even many ardent cereologists admit there are more hoaxed than "genuine" ones (Nickell 1996a; Nickell and Fischer 1992). For example, prominent cereologist Colin Andrews (2001) has conceded that 80 percent of the British crop circles are manmade, yet Levengood claims his research "suggests that over 95% of worldwide crop formations involve organized ion plasma vortices" (Levengood and Talbott 1999).

Levengood and others who postulate crop-stamping, ion plasma vortices have to face the fact that those phenomena/entities remain unrecognized by science. They owe their imagined existence to George Terence Meaden, a former professor of physics who took up meteorology as an avocation. His book, *The Circles Effect and Its Mysteries* (1989), is still revered by many cereologists. Alas, however, he merely attempted to "explain" a mystery by creating another, and—humiliated by hoaxers—eventually retired from the scene, *conceding that all of the complex designs were fakes* (Hoggart and Hutchinson 1995, 59).

Nevertheless, many circles aficionados have begun to photograph supposed vortex effects which, curiously, resemble some of the same photographic anomalies that are the stock-in-trade of ghost hunters. For example, in her *Mysterious Lights and Crop Circles,* credulous journalist Linda Moulton Howe (2000, 137, 255) exhibits a flash photograph taken in a crop circle that shows a bright "mysterious arch with internal structure that seems to spiral like a plasma." Unfortunately for Howe (erstwhile promoter of cattle mutilations and similar "mysteries"), the effect is indistinguishable from that caused by the camera's unsecured wrist strap reflecting the flash (Nickell 1996b). As corroborative evidence of this mundane cause, the bright strand-like shapes typically go unseen by the ghost hunter or cereologist, appearing only in their snapshots.

Again, Howe shows several photos containing "transparent spheres" that the croppies variously call "energy balls," "light orbs," "atmospheric plasmas," and so on (2000, 169–76). They are indistinguishable from "orbs" of "spirit energy" typically seen in photographs of graveyards and other "haunted" places and that sometimes appear in snapshots as UFOs. Skeptics have demonstrated that these globelike effects can be produced by particles of dust, water droplets, and the like reflecting the flash (Mosbleck 1988; Nickell 1994; Burton 1999). Other simulators of paranormal "energy" in photos include lens flares (the result of interreflection between lens surfaces), bugs and debris reflecting the flash, and many other causes, including film defects and outright hoaxes (Nickell 1994).

Sometimes, however, "hovering balls of light" and other "energy" effects are reported by eyewitnesses, though not only in the vicinity of crop circles (Haselhoff 2001; Howe 2000). These too may have a variety of causes, including pranksters' parachute flares ("Flares" 1999), various misperceived aerial craft and other phenomena (such as ball lightning), false claims, hallucinations, and others. In some instances, small lights observed moving about cropfields at night might have come from the flashlights of the circlemakers!

It appears that for the foreseeable future, the crop-circle phenomenon will continue. At least it has moved from the level of mere hoaxing—"a form of graffiti on the blank wall of southern England" (Johnson 1991)—to represent an impressive genre of outdoor art. The often breathtaking designs (best seen in aerial photographs, like the giant Nazca drawings in Perú) are appreciated not only by the mystery mongers but by skeptics as well. In fact, as reliably reported (Hoggart and Hutchinson 1995), skeptics have helped to make many of them!

REFERENCES

Andrews, Colin. 2001. Cited in *The Deepening Complexity of Crop Circles,* edited by Eltjo H. Haselhoff, 37–38. Berkeley, Cal.: Frog, Ltd.

Burton, Garry. 1999. Welcome to "Orb World." Retrieved from http://members.aol.com/Analogsys/index2.html.

Delgado, Pat, and Colin Andrews. 1989. *Circular Evidence.* Grand Rapids, Mich.: Phanes Press.

Flares spark reports of UFOs. 1999. *Cornish Guardian*, 19 August (quoted in Howe 2000, 166–67).

Haselhoff, Eltjo H. 2001. *The Deepening Complexity of Crop Circles.* Berkeley, Cal.: Frog, Ltd.

Hoggart, Simon, and Mike Hutchinson. 1995. *Bizarre Beliefs.* London: Richard Cohen Books, 52–61.

Howe, Linda Moulton. 2000. *Mysterious Lights and Crop Circles.* New Orleans, La.: Paper Chase Press.

Johnson, Jerold R. 1991. Pretty pictures. *MUFON UFO Journal* 275, 18 March.

Levengood, W. C., and Nancy P. Talbott. 1999. Dispersion of energies in worldwide crop formations. *Physiologia Plantarum* 105: 615–24.

Meaden, George Terence. 1989. *The Circles Effect and Its Mystery.* Bradford-on-Avon, Wiltshire: Artetech.

Mosbleck, Gerald. 1988. The elusive photographic evidence. In *Phenomenon: Forty Years of Flying Saucers,* edited by John Spencer and Hilary Evans, 210. New York: Avon.

Nickell, Joe. 1994. *Camera Clues: A Handbook for Photographic Investigation.* Lexington, Ky.: University Press of Kentucky.

———. 1995. Crop circle mania wanes. *Skeptical Inquirer* 19, no. 3 (May/June): 41–43.

———. 1996a. Levengood's crop-circle plant research. *Skeptical Briefs* 6, no. 2 (June): 1–2.

———. 1996b. Ghostly photos. *Skeptical Inquirer* 20, no. 4 (July/August): 13–14.

Nickell, Joe, and John F. Fischer. 1992. The crop-circle phenomenon: An investigative report. *Skeptical Inquirer* 16, no. 2 (Winter): 136–49.

Spencer, John, and Hilary Evans, eds. 1988. *Phenomenon: Forty Years of Flying Saucers.* New York: Avon.

15

Zanzibar Demon

The scene is modern-day Zanzibar, where a terrible monster, the infamous "popobawa," is swooping into bedrooms at night and raping men—particularly skeptical men. The demonic beast's name comes from the Swahili words for *bat* and *wing*, and indeed the creature is described as having, in addition to a dwarf's body with a single cyclopean eye, small pointed ears, and batlike wings and talons. According to local villagers, it is especially prone to attack "anybody who doesn't believe" (McGreal 1995).

One 1995 victim was a quiet-spoken peasant, a farmer named Mjaka Hamad, who said he does not believe in spirits. He first thought he was having a dream. However, "I could feel it," he said, "something pressing on me. I couldn't imagine what sort of thing was happening to me. You feel as if you are screaming with no voice." He went on to say: "It was just like a dream but then I was thinking it was this popobawa and he had come to do something terrible to me, something sexual. It is worse than what he does to women."

The demon struck Zanzibar in 1970 and again briefly in the 1980s. According to *The Guardian,* "Even those who dismiss the attacks as superstition nonetheless admit that for true believers they are real. Zanzibar's main hospital has treated men with bruises, broken ribs and other injuries, which the victims blame on the creature" (McGreal 1995).

I was given an article on the Zanzibaran affair by a colleague who half-jokingly remarked, "Here's a case for you to solve." I read a few paragraphs and replied, "I have solved it."

I only needed to recall some of my earlier research to realize that the popobawa is simply a Zanzibaran version of a physiological and psychological phenomenon known as a "waking dream." One of the characteristics of such a dream, known more technically as a *hypnopompic* or *hypnagogic* hallucination (depending on whether one is, respectively, waking up or going to sleep), is a feeling of being weighted down or even paralyzed. Alternately, one may "float" or have an out-of-body experience. Other characteristics include extreme vividness of the dream and bizarre and/or terrifying content (Baker and Nickell 1992, 226–27; Nickell 1995, 41, 46, 55, 59, 117, 131, 157, 209, 214, 268, 278).

Similar feelings were also experienced by persons in the Middle Ages who reported nighttime visitations of an *incubus* (a male demon that lay with women) or a *succubus* (which took female form and lay with men). In Newfoundland the visitor was called the "Old Hag" (Ellis 1988). In the infamous West Pittston, Pennsylvania, "haunted house" case of 1986, tenant Jack Smurl claimed he was raped by a succubus. As "demonologist" Ed Warren described it:

> He was asleep in bed one night and he was awakened by this haglike woman who paralyzed him. He wanted to scream out, of course—he was horrified by what he saw, the woman had scales on her skin and white, scraggly hair,

> and some of her teeth were missing—but she paralyzed him in some manner. Then she mounted him and rode him to her sexual climax. [Warren and Warren 1989, 105–6].

Such accounts come from widespread places and times. For example, consider this interesting encounter, which occurred in the seventeenth century. It concerned one Anne Jeffries, a country girl from Cornwall. According to Ellis (1988):

> In 1645 she apparently suffered a convulsion and was found, semi-conscious, lying on the floor. As she recovered, she began to recall in detail how she was accosted by a group of six little men. Paralyzed, she felt them swarm over her, kissing her, until she felt a sharp pricking sensation. Blinded, she found herself flying through the air to a palace filled with people. There, one of the men (now her size) seduced her, and suddenly an angry crowd burst in on them and she was again blinded and levitated. She then found herself lying on the floor surrounded by her friends [264].

This account obviously has striking similarities to many UFO abduction accounts—some of which, like those of Whitley Strieber's own "abduction" experiences described in *Communion* (1988), are fully consistent with hypnopompic or hypnagogic hallucinations (Baker and Nickell 1992). Still other entities that have appeared in classic waking dreams are ghosts and angelic visitors (Nickell 1995).

As these examples illustrate, although the popobawa seems at first a unique, Zanzibaran creature, it is actually only a variant of a well-known phenomenon—one that Western skeptics, at least, have little to fear.

REFERENCES

Baker, Robert A., and Joe Nickell. 1992. *Missing Pieces: How to Investigate Ghosts, UFOs, Psychics and Other Mysteries.* Buffalo, N.Y.: Prometheus Books.

Ellis, Bill. 1988. The varieties of alien experience. *Skeptical Inquirer* 12, no. 3 (Spring): 263–69.

McGreal, Chris. 1995. Zanzibar diary. *The Guardian,* 2 October.

Nickell, Joe. 1995. *Entities: Angels, Spirits, Demons, and Other Alien Beings.* Amherst, N.Y.: Prometheus Books.

Streiber, Whitley. 1988. *Communion: A True Story.* New York: Avon.

Warren, Ed, and Lorraine Warren, with Robert David Chase. 1989. *Ghost Hunters.* New York: St. Martin's Paperbacks.

16

Winchester Mystery House

It is the work of an eccentric widow, who was supposedly guided by spirits and a construction project that lasted 38 years. It began in 1884 with an existing but unfinished 8-room farmhouse near San Jose, California, and culminated in a 7-story, turreted, Gothic Victorian mansion that once comprised an estimated 500 rooms. The 1906 San Francisco earthquake significantly reduced the stacked and sprawling architectural wonder, but when Sarah Winchester died in 1922 it still "contained 160 rooms, 2,000 doors, 10,000 windows, 47 stairways, 47 fireplaces, 13 bathrooms, and 6 kitchens" (*Winchester* 1997).

Even more remarkable, the round-the-clock construction yielded "an interminable labyrinth" of "miles of twisting hallways" (*Winchester* 1997, 14). Indeed, the house became "an architectural nightmare," featuring rooms with odd angles, stairways leading to nowhere, secret passageways, doors and windows opening onto blank walls, and railing posts set upside down (Guiley 2000, 405–407; Murray 1998, 57–66).

FIGURE 16-1. This view of the Winchester Mystery House, with the author standing in front, fails to convey the immensity of the mansion.

FIGURE 16-2. View from a window shows part of the sprawling grounds, including the bell tower that, allegedly, was once used to summon spirits. (Photographs by Joe Nickell.)

Fascinating in its own right, Sarah Winchester's remarkable story has been embellished—rather like her strange mansion itself—with implausible incidents, ornate details, and "facts" that lead, tortuously, to dead ends. The truth is elusive because she was never interviewed and left no diary or other written record. Moreover, "wild stories" about the lady proliferated in her lifetime as well as after her death, and popular writers garbled—and invented—details to suit their own purposes (*Winchester* 1997, 11).

In an attempt to sort truth from fiction, I toured Winchester Mystery House (accompanied by Vaughn Rees of the Center for Inquiry–West) on October 24, 2001, as part of a California speaking and investigations tour. Subsequently I delved into many of the books and other sources of lore and legend about Sarah Winchester and her curious obsession. Here is some of what I found.

- **Fancy:** The story begins in 1862 in New Haven, Connecticut, with Sarah L. Pardee's marriage to William Wirt Winchester. He was the son and heir of Oliver Fisher Winchester, whose repeating firearm became famous as "The Gun That Won the West" (*Winchester* 1997, 46). However, according to one tale, "[d]uring a Connecticut thunder storm, Mrs. Winchester's husband and baby lost their lives in a tragic fire" (quoted in Rambo 1967, 6). But wait: Another source (Smith 1967, 35–43) states, "When tragedy struck this woman, it pulled no punches. Her husband died a lingering death from tuberculosis, and her little girl passed away almost immediately afterward." In yet another source, *The National Directory of Haunted Places*, Hauck (1996, 75–76) separates the deaths with a bit more time, stating that "her husband and only child died within months of each other."
  Fact: Actually, the Winchesters' infant daughter Annie died first, in 1866. It was not until 1881, *15 years later*, that Sarah's husband passed away (*Winchester* 1997, 8).
- **Fancy:** Mrs. Winchester, grief-stricken but having inherited her husband's fortune, sought out a Boston medium to contact his spirit. The medium, Adam Coons, relayed the message, somehow rediscovered by Susy Smith (1967): "This is a warning. You will be haunted forever by the ghosts of those who have been killed by Winchester rifles, unless you make amends to them." She was instructed to head west, guided by her husband's spirit.

  However, another source states that it was not a male medium but "a seeress" who provided Mrs. Winchester with the message (again, somehow rediscovered): "There is a curse on your life.

It is the same curse that took your child and husband . . . that has resulted from the terrible weapon that the Winchester family created. . . . [B]uild a house not only for yourself but also for the spirits of those who have fallen before that terrible weapon. As long as you build, you will live. Stop and you will die" (Winer and Osborn 1979, 33–49).

Fact: It has never been proven that Sarah Winchester ever consulted a "Boston medium," whether male or female. One local historical writer maintains that after her husband's death, Mrs. Winchester was indeed grief-stricken. "Doctors and friends urged her to leave the East, seek a milder climate and search for some all consuming hobby. One physician *did* suggest that she 'build a house and don't employ an architect.'" That researcher concedes, "Perhaps she *was* a spiritualist," but insists, "Miss Henrietta Severs, her constant companion for years, always firmly denied she had any Spiritualist leanings" (Rambo 1967, 8).

▶ **Fancy:** Sarah Winchester's "curious building techniques" resulted from her desire "to control the evil entities and keep them from harming her." For example, "[o]ne stairway, constructed like a maze, has seven flights and requires forty-four steps to go ten feet" (Smith 1967, 38). Some interior rooms have barred windows, the floor in one room is comprised of trap doors, and there are doors and stairs that lead nowhere (Rambo 1967; Murray 1998, 59).

Fact: The winding stair, with its two-inch-high steps, had nothing to do with ghosts and everything to do with Mrs. Winchester's severe arthritis and neuritis. The low steps were built to accommodate her diminished abilities (just as elevators were later installed when she was forced to use a wheelchair). There is an equally simple explanation for the curiously barred interior windows: they were once *exterior* windows, but the constant additions to the house relegated them to the inside. The doors and stairs that lead

FIGURE 16-3. "Switchback" staircase takes 44 steps and 7 turns to advance less than 10 feet upward, and was supposedly created to baffle spirits. The wall at the right, though, shows the silhouette of the original steps; they were replaced to accommodate Mrs. Winchester's debilities. (Photograph by Joe Nickell.)

to dead ends are similarly explained. As to the floor with trap doors, those are in a special greenhouse room; they were designed to open onto a zinc subfloor so that runoff from watered plants could be drained by pipes to the garden beneath (Rambo 1967; *Winchester* 1997; Palomo 2001).

- **Fancy:** Sarah Winchester's blue séance room, her "secret rendezvous with the spirits," was off limits to her huge staff of carpenters and servants. There, at midnight—while a bell in a tower was rung to summon spirits—she donned one of 13 ceremonial robes she wore when communing with the entities. She also held parties for the spirits, offering them caviar, stuffed pheasant, and other dishes served on gold plates which she otherwise kept in her safe. Sometimes, late at night, "ghostly music" was heard "wafting from the dark mansion" (*Winchester* 1997, Rambo 1967, Winer and Osborn 1979).

Fact: According to historical writer Ralph Rambo, whose father had helped with the landscaping of the mansion grounds, the allegedly sacrosanct "séance" room "was also utilized as a bedroom at various times by Mrs. Winchester's foreman, the chauffeur, the head Japanese gardener and his wife" (1967, 8–9). In addition, the bell was used, not for midnight spirit assembly, but as a call to and from work and as an alarm in case of fire. There is no evidence that Mrs. Winchester had 13 ceremonial robes; that fiction was probably based on the total of 13 hooks in the room's two closets, though some of them are placed implausibly low for robes. When the safe was opened after her death, there was no solid-gold dinner service, only mementos, including a lock of her baby's hair. Acknowledging the "legend" of the plates that were supposedly used to serve exotic dishes to phantom guests, a Winchester Mystery House publication (*Winchester* 1997) states: "On the other hand, this theory might have come from rumors about the mansion's well-fed servants!" The "ghostly music" has an even simpler explanation: When she was unable to sleep, Mrs. Winchester often played the pump organ in the Grand Ballroom.

- **Fancy:** Mrs. Winchester was so reclusive that she always wore a black veil. She also refused admission to most visitors, including Mary Baker Eddy, the founder of Christian Science. President Theodore Roosevelt, a fan of Winchester rifles, knocked at the front door but was not recognized by a servant and hence was turned away (or insulted by being sent around to the back of the house) (*Winchester* 1997; Rambo 1967). "One of her few guests was Harry Houdini, who never spoke of his single visit to Winchester House" (Hauck 1996).

  Fact: Although Mrs. Winchester was reclusive, the single existing photograph taken of her during the 38 years of mansion construction shows her seated in her carriage, apparently gazing at the camera, without any veil (*Winchester* 1997, 6). There are various versions of the Teddy Roosevelt story, but Ralph Rambo (1976, 9)—who was "standing directly across the road that day"—says the president was merely driven past the house. The local Chamber of Commerce had asked permission for Mr. Roosevelt to visit, but received Mrs. Winchester's sharp "No!" Whether Mary Baker Eddy gained entrance is uncertain (Hauck 1996; Smith 1967, 40; Rambo 1967). However, Houdini *did* visit the house and *was* admitted—even joining in a midnight séance!—but this took place in November 1924, two years after Sarah Winchester's death. Houdini discussed his visit in an article in Portland's *Oregon Daily Journal* (*Winchester* 1997, 42).

- **Fancies:** In the early years of her residence, Mrs. Winchester planned a lavish reception for Santa Clara Valley residents, sending out hundreds of gilded engravings. She had servants prepare "a sumptuous midnight banquet" and hired "a famous orchestra" for entertainment, but not a single guest appeared. In 1906 the Great San Francisco Bay Area earthquake toppled the seven-story tower onto Mrs. Winchester's bedroom where she was trapped for several hours, moaning, "Oh, God help me. The evil spirits have taken over the house." After servants rescued her, the terrified widow fled to Redwood City and did not return for six years (Rambo 1967; Smith 1967, 41–42).

Facts: The tale of Sarah Winchester's grandiose but unattended reception for local citizens is "pure, unadulterated balderdash!" Although she was indeed trapped by the earthquake and subsequently relocated, her absence was only for a period of about six months while the house was partially repaired, not six years (Rambo 1967).

▶ **Fancy:** The Winchester Mystery House is the "most haunted house in the country" ("Global" 2002), being tenanted with "thousands of ghosts and guests" (Harter 1976, 54–61) as well as "the spirit of Sarah Winchester herself" (Murray 1998, 63). According to *The Encyclopedia of Ghosts and Spirits* (Guiley 2000, 405–7):

> Many visitors are haunted by various phenomena, such as phantom footsteps, odd sounds, eerie quiet, whisperings, sounds of a piano playing, smells of phantom food cooking, cold spots, doorknobs turning by themselves, and windows and doors slamming shut. The floor of the gift shop has been found mysteriously covered with water and items in disarray.

One guide characterized the Daisy Bedroom, where Mrs. Winchester was trapped by the earthquake in 1906, as the room that most frightened him. When ghost hunters Winer and Osborn (1979, 43) asked why:

> "I can't really pinpoint any one thing," replied the guide. "But sometimes that room gets so chilly. Not the whole

> room but just in certain parts. And there's that feeling when I'm in there alone like maybe I'm being watched, like I'm not alone. Some of the others told me that they get the same feeling in that room."

> **Fact:** There is no scientific evidence that the Winchester Mystery House—or any place—is haunted. As psychologist Robert A. Baker is fond of saying, tongue in cheek, there are instead "only haunted people." The Winchester hauntees often report mere feelings, like the guide frightened by the Daisy Bedroom. I had no such feelings when I lingered in that room; in any case, they can be the products of imagination provoked by suggestion. This in turn can be attributed to the mansion's gothic ambiance and the many legends of ghosts, which create a certain expectancy in many people.

As well, it would be unusual if such a rambling old house did *not* have drafts, temperature variations and fluctuations, and odd noises caused by changing temperatures, the settling of an old structure, and other causes, including seismic activity. "Whisperings" can easily be imagined, or can be the product of wind or other causes. The sounds of ghostly music can similarly be imaginary. In at least one instance, "piano music" was "heard" in the house by one person but not by her companion (Winer and Osborn 1979, 46–47); on another occasion, a "psychic" claimed to hear "organ music" although others who were present did not (Myers 1986). Actual music may even be perceived, as from a radio in a passing car. So bent on fostering mystery are some that, when odd noises are not forthcoming, the *absence* of them—an "eerie quiet"—will do.

In one instance, "a shadowy figure emerged from the inner recesses of the huge mystery house," but turned out to be a Winchester staffer (Winer and Osborn 1979, 46). A tour group saw an elderly woman sitting at a kitchen table, which the guide only later decided was a ghost (Myers 1986, 48); more likely, she was just a straggling member of an-

FIGURE 16-4. **Investigator Vaughn Rees points to damage from the 1906 earthquake in the Daisy Bedroom. The room's austere appearance may help foster the creepy feelings that some claim to experience there. (Photograph by Joe Nickell.)**

other tour taking a brief rest. (Whether the woman was really "dressed like" Mrs. Winchester, as later recalled, could be a misremembering caused by suggestion.) The occasional apparition may be nothing more than the welling-up of a mental image deriving from a daydream or other altered state of consciousness, which is then superimposed upon the current visual scene. This phenomenon is especially common to those who have traits associated with a fantasy-prone personality (Nickell 2000)—like "psychic" Sylvia Browne who described being watched by two spirits from across a room (*Winchester* 1997, 42).

A caretaker's being awakened one night by an *unlikely* sound—that "of a screw being unscrewed, then hitting the floor and bouncing onto a carpet runner" (*Winchester* 1997, 42)—might be due to a "waking dream." This especially real-seeming occurrence is actually a common type of hallucination that takes place in the twilight between wakefulness and sleep (Nickell 1995). Additional reported ghostly or odd occurrences may have still other mundane explanations, such as reported food smells (wafting from the gift-shop restaurant?), "mysterious moving lights" (reflections, as from one of the mansion's art-glass windows framed with glittering "jewels"?), a water-soaked floor (a leak, or spill or condensation or prank or even vandalism, among other many other possible causes), and so forth (*Winchester* 1997, 19; Murray 1998; Myers 1986, 45–51).

Once the idea that a place is haunted takes root, almost any unknown noise, mechanical glitch, or other odd occurrence can become added "evidence" of ghostly shenanigans, at least to susceptible people. They often cite "unexplain*able*" phenomena, but they really mean "unexplain*ed*," a condition that does not in any way imply or necessitate the supernatural. To suggest that it does is to engage in a logical fallacy called arguing from ignorance—the stock-in-trade of credulous paranormalists and outright mystery-mongering writers.

Not everyone is susceptible. My docent, a senior tour guide at the Winchester Mystery House (Palomo 2001), has had no experiences herself, although, if others are to be believed, she has had endless opportunities. From three decades of ghost investigating, I have noted a pattern suggesting that as the level of individuals' ghostly experiences rises, so does their propensity for imagination and fantasy (Nickell 2000)—evidence, it seems, for "haunted people."

REFERENCES

Global Halloween Alliance. 2002. *Happy Halloween Magazine*, vol. 3, no. 3: http://www.halloweenalliance.com/magazine/vol3iss3_wmh.htm

Guiley, Rosemary Ellen. 2000. *The Encyclopedia of Ghosts and Spirits*. 2d ed. New York: Checkmark Books.
Harter, Walter. 1976. *The Phantom Hand and Other American Hauntings*. Englewood Cliffs, N.J.: Prentice-Hall.
Hauck, Dennis William. 1996. *The International Directory of Haunted Places*. New York: Penguin Books.
Murray, Earl. 1998. *Ghosts of the Old West*. New York: Tom Doherty Associates.
Myers, Arthur. 1986. *The Ghostly Register*. Chicago: Contemporary Books.
Nickell, Joe. 1995. *Entities*. Amherst, N.Y.: Prometheus Books.
———. 2000. Haunted inns. *Skeptical Inquirer* 24, no. 5 (September/October): 17–21.
Palomo, Cathy. 2001. Senior Tour Guide, Winchester Mystery House. Personal communication, 24 October.
Rambo, Ralph. 1967. *Lady of Mystery*. San Jose, Cal.: The Press.
Smith, Susy. 1967. *Prominent American Ghosts*. New York: World Publishing Co.
The Winchester Mystery House. 1997. San Jose, Cal.: Winchester Mystery House.
Winer, Richard, and Nancy Osborn. 1979. *Haunted Houses*. New York: Bantam Books.

17

Voodoo in New Orleans

New Orleans has been declared America's most haunted city (Klein 1999, 104), and tour guides—following the imaginative lead of Anne Rice—have attempted to overlay its rich history with bogus legends of vampires and other spine-tingling notions. But perhaps the city's oldest and most profound occult traditions are those involving the mysterious practices of voodoo. During a southern speaking tour, I was able to set aside a few days to explore the New Orleans museums, shops, temples, and tombs that relate to this distinctive admixture of religion and magic, commerce and controversy.

Voodoo

Voodoo—or *voudou*—is the Haitian folk religion. It consists of various African magical beliefs and rites that have become mixed with Catholic elements. It began with the arrival of slaves in the New World, most of them from the western, "Slave Coast" area of Africa, notably Dahomey (now Benin) and Nigeria. In Benin's Fon language, *vodun* means "spirit," an invisible, mysterious force that can intervene in human affairs (Hurbon 1995, 13; Métraux 1972, 25, 359; Bourguignon 1993).

According to one writer, "The Blacks suffered under merciless circumstances—their property and their family and social structures all torn to shreds; they had nothing left—except their Gods to whom they clung tenaciously." In Haiti and elsewhere, there was an attempt to strip them even of that, as their "heathen" beliefs were rigorously suppressed. However, the slaves "worshipped many of their Gods unbeknownst to the priests, under the guise of worshipping Catholic saints" (Antippas 1988, 2).

Voodoo's African elements include worship of *loa* (supernatural entities) and the ancestral dead, together with the use of drums and dancing, during which loa may possess the faithful. Catholic elements include prayers such as the Hail Mary and the Lord's Prayer, as well as rituals such as baptism, making the sign of the cross, and the use of candles, bells, crosses, and images of saints. Many of the *loa* are equated with specific saints. For example, Damballah, the Dahomean snake deity, is identified with St. Patrick who, having legendarily expelled all snakes from Ireland, is frequently depicted stamping on snakes or brandishing his staff at them (Bourguignon 1993).

Voodoo spread from Haiti to New Orleans in the wake of the Haitian slave revolt (1791–1804). The refugee plantation owners fled with their slave retinues to Louisiana, where slaves had previously toiled under such repressive circumstances that their African religion "had all but withered." However, oppression lessened somewhat with American rule, following the Louisiana Purchase of 1803, and—with the influx of thousands of voodoo practitioners—soon "New Orleans began to hear the beat of the drum" (Antippas 1988, 14).

Voodoo Queen

Voodoo in New Orleans can scarcely be separated from its dominant figure, Marie Laveau, about whom many legends swirl. According to one source (Hauck 1996, 192, 193):

> She led voodoo dances in Congo Square and sold charms and potions from her home in the 1830s. Sixty years later she was still holding ceremonies and looked as young as she did when she started. Her rites at St. John's Bayou on the banks of Lake Pon[t]chartrain resembled a scene from hell, with bonfires, naked dancing, orgies, and animal sacrifices. She had a strange power over police and judges and succeeded in saving several criminals from hanging.

Writer Charles Gandolfo (1992), author of *Marie Laveau of New Orleans,* stated: "Some believe that Marie had a mysterious birth, in the sense that she may have come from the spirits or as an envoy from the Saints." In contrast, a plaque on her supposed tomb, placed by the Catholic Church, refers to her as "this notorious 'voodoo queen.'"

Who was the real Marie Laveau? She began life as the illegitimate daughter of a rich Creole plantation owner, Charles Laveaux, and his Haitian slave mistress. Sources conflict, but Marie was apparently born in New Orleans on September 10, 1801 (Jensen 2002). In 1819 she wed Jacques Paris who, like her, was a free person of color, but she was soon abandoned or widowed. About 1826, she began a second, common-law marriage to Christophe de Glapion, another free person of color.[1]

Marie was introduced to voodoo by various "voodoo doctors," practitioners of a popularized voodoo that emphasized curative and occult magic and seemed to have a decidedly commercial aspect. Her own practice began when she teamed up with a "heavily tattooed Voodoo doctor"—known variously as Doctor John, Bayou John, John Bayou, etc.—who was "the first commercial Voodooist in New Orleans to whip up potions and gris-gris for a price" (Gandolfo 1992, 11). *Gris-gris* or *juju* refers to magic charms or spells, which often took the form of conjuring bags containing such items as bones, herbs, charms, snake skin, and so on tied up in a piece of cloth (Antippas 1988, 16). Doctor John reportedly confessed to friends that his magic was mere humbuggery. "He had been known to laugh," wrote Robert Tallant, in *Voodoo in New Orleans* (1946, 39), "when he told of selling a gullible white woman a small jar of starch and water for five dollars."

In time Marie decided to seek her own fortune. Working as a hairdresser, which put her in contact with New Orleans' social elite, she soon developed a clientele to whom she dispensed potions, gris-gris bags, voodoo dolls, and other magical items. She then sought supremacy over her rivals, some 15 "voodoo queens" in various neighborhoods. According to a biographer (Gandolfo 1992, 17):

> Marie began her take-over process by disposing of her rival queens. . . . If her rituals or gris-gris didn't work, Marie (who was a statuesque woman, to say the least) met them in the street and physically beat them. This war for supremacy lasted several years until, one by one, all of the former queens, under a pledge, agreed to be sub-queens. If they refused, she ran them out of town.

By the age of 35, Marie Laveau had become New Orleans's most powerful voodoo queen—then or since. She won the approval of the local priest by encouraging her followers to attend Mass. Although she charged the rich abundantly, she reportedly gave to the needy and administered to the suffering. Her most visible activities, however, were her public rituals. By municipal decree (from 1817), slaves were permitted to dance publicly only at a site called Congo Square. "These public displays

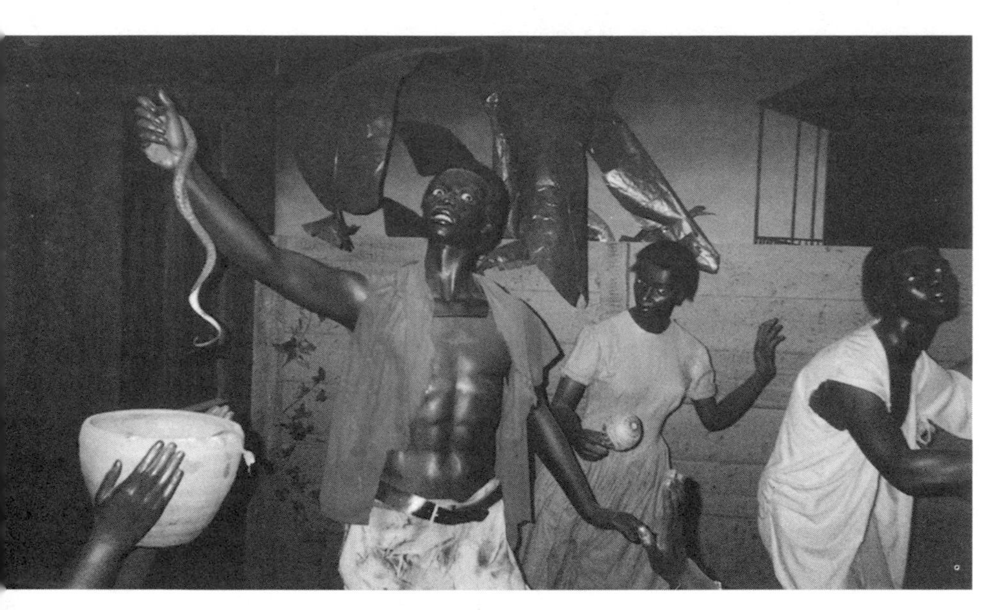

FIGURE 17-1. **Tableau, "Marie Laveau's Voodoo Dancers," in the Musée Conti wax museum.**

of Voodoo ceremonies, however, revealed nothing of the real religion and became merely entertainment for the curious whites" (Antippas 1988, 14–15). More "secret" rituals—including fertility rituals—took place elsewhere, notably on the shore of Lake Pontchartrain.

It is difficult at this remove to assess just how much of Marie's rituals was authentic voodoo practice and how much was due to her "incredible imagination and an obsession for the extreme." She staged rituals that were "simulated orgies." Men and women danced with abandon after drinking rum and seeming to become possessed by various loas (Figure 17-1). Seated on her throne, Marie directed the action when she was not actually participating. She kept a large snake, called Le Grand Zombi, with which she would dance in veneration of Damballah, shaking a gourd rattle to summon that snake deity and repeating over and over, "Damballah, ye-ye-ye!"

Once a year Marie presided over the ritual of St. John's Eve. It began at dusk on June 23 and ended at dawn on the next day, St. John's Day. Hundreds attended, including reporters and curious onlookers, each of whom was charged a fee. Drum beating, bonfires, animal sacrifice, and other elements—including nude women dancing seductively—characterized the extended ritual. Offerings were made to the appropriate loas for protection, including safeguarding children and others from the Cajun bogeyman, the *loup-garou*, a werewolf that supposedly fed on the blood of victims (Gandolfo 1992, 18–23).

Magic or Myth?

Claims regarding Marie Laveau's alleged powers persist. She represented herself as a seer and used such fortune-telling techniques as palmistry (Gandolfo 1992, 26). There is, however, no evidence that Marie's clairvoyant abilities were any more real or successful than those of any other fortuneteller. We know that people attest to the accuracy of such readings because they do not understand the clever techniques involved, like "cold reading." So called because it is accomplished without any foreknowledge, this is an artful method of fishing for information from the sitter while convincing him or her that it comes from a mystical source (Hyman 1977).

Actually, many of Marie's readings may not have been so "cold" after all. Far from lacking prior information about her clients, she reputedly used her position as a hairdresser for gossip collecting, discovering "that her women clients would talk to her about anything and everything and would divulge some of their most personal secrets to her" (Gandolfo 1992, 12). She also reputedly "developed a chain of household informants in most of the prominent homes" (Antippas 1988, 16).

Doubtless such intelligence gathering would be helpful to a fortune-telling enterprise (just as "mediumistic espionage" was utilized by later spiritualists [Keene 1976, 27]). It could also be beneficial to a business, like Marie's, that dispensed charms:

> Most of her work for the ladies involved love predicaments. Marie knew the personal secrets of judges, priests, lawyers, doctors, ship captains, architects, military officers, politicians, and most of New Orleans' other leading citizens. She used her knowledge of their indiscretions and blackmailed them into doing whatever she wanted. She was then financially reimbursed by her elite female clients. Most of the time, this was how her love potions and gris-gris worked, which is apparently 100% of the time [Gandolfo 1992, 12].

Such tactics may help explain the claim, mentioned earlier, that Marie "had a strange power over police and judges and succeeded in saving several criminals from hanging" (Hauck 1996, 192, 193). Still, we should beware of taking such claims too seriously. When we seek to learn the facts, we soon realize that we have entered the realm of folklore. There are, for example, rather conflicting versions of one case, circa 1830, in which Marie performed certain rituals, at the request of his father, on behalf of an unidentified young man who was charged with "a crime" (rape, according to one source). Supposedly, either the case was dismissed or the young man was acquitted, and Marie was rewarded with a cottage on Rue St. Ann. However, as one writer conceded, "No one is sure how Marie actually won the case." Therefore, of course, there is no evidence that she *did* (Gandolfo 1992, 14–15; *cf.* Tallant 1946, 58; Martinez 1956, 17–19).

Legends of Marie's beneficent aspect are rivaled by those of her sinister one. A story in this regard involves her alleged hex on a New Orleans businessman, J. B. Langrast, in the 1850s. Langrast supposedly provoked Marie's ire by publicly denouncing her and accusing her of everything from robbery to murder. Soon, gris-gris in the form of roosters' heads began to appear on his doorstep. As a consequence, Langrast reportedly grew increasingly upset and eventually fled New Orleans (Nardo and Belgum 1991, 89–92).

I have traced the Langrast story to a 1956 book of Mississippi folk-tales, which describes the "businessman" as a junk dealer and bigamist (Martinez 1956, 78–83). Such a man might have various reasons for leaving town. Claims that Marie Laveau invoked a loa to curse Langrast with insanity are invalidated by a complete lack of proof that he ever became insane. In fact, his alleged flight could easily be attributed to simple fear, the belief that "Marie Laveau's followers might kill him if he stayed" (Nardo and Belgum 1991, 90–91).

Marie II

Among the most fabulous legends about Marie Laveau is the often-repeated one alleging "her perpetual youth" (Hauck 1996). According to a segment of "America's Haunted Houses" (1998), which aired on the Discovery Channel, Marie was "said to be over 100 years old when she died and as beautiful as ever." Moreover, "[t]here were some unexplained and mysterious sightings of the great Voodoo Queen even after her death," writes Gandolfo (1992, 29). "People would swear on a stack of bibles that they saw Marie Laveau herself." Indeed, he adds, "A number of people say they were at a ritual in the summer of 1919 given by the Great Queen."

The solution to this enigma is the fact that, according to Tallant (1946, 52), there were "at least two Marie Laveaus." The first Marie, the subject of our previous discussions, died June 15, 1881. Her obituaries say she was then 98 ("Marie Lavaux" 1881; "Death" 1881). However, the doctor who attended Marie at the end publicly stated his doubts that she was as old as her family claimed; he judged her age to be in the late eighties (Tallant 1946, 117). In fact, she was not quite 80.

Whatever her actual age, far from being a figure of eternal youth, Marie Laveau spent her last years "old and shrunken," reportedly stripped of her memory, and lying in a back room of her cottage (Tallant 1946, 88, 115). In her stead was her daughter, Marie Laveau II. The younger Marie gradually took over her mother's business activities, which included running a house on Lake Pontchartrain where rich Creole men could have "appointments" with young mulatto girls (Tallant 1946, 65–66). She died in 1897. The claim that Marie Laveau was active in 1919 is thought to have been based on a third Marie, possibly a granddaughter (Gandolfo 1992, 29), or another voodoo queen with whom she was confused.

In carrying on her mother's work, Marie II had business cards printed, billing her not as a voodooienne but as a "Healer." According to Tallant (1946, 93):

> The Laveau ways of performing homeopathic magic were endless. Sick people were often brought to the house to receive the benefit of a cure by Marie II. A person bitten by a snake was told to get another live snake of any sort, cut its head off "while it was angry" and to tie this head to the wound. This was to be left attached until sunrise of the following day. Sometimes her practices contained an element of medical truth, embracing the use of roots and herbs that contained genuine curative elements. For sprains and swellings she used hot water containing Epsom salts and rubbed the injured parts with whiskey, chanting prayers and burning candles at the same time, of course. For other ailments she administered castor oil, to the accompaniment of incantations and prayer.

Like other occult healers, Marie obviously took advantage not only of the occasional "element of medical truth," but also of other factors, including the body's own natural healing mechanisms and the powerful effects of suggestion.

Voodoo Today

Current voodoo practice in New Orleans is a mere shadow of what it was in its heyday. Although an estimated 15 percent of the city's population supposedly still practices voodoo, it has largely been subsumed into Catholicism, which remains the dominant religion. It has also been influenced by spiritualist, Wiccan, and other occult and New-Age beliefs (Gandolfo n.d.). The most visible aspects of voodoo today are tours and attractions in the area of the Vieux Carré (or "Old Square"), popularly called the French Quarter. Laid out in 1721, it is the oldest area of the city.

There, souvenir shops sell that most stereotypical of items associated with voodoo, the voodoo doll. Although in the days of Marie Laveau one might occasionally encounter "a little wax doll stuck with pins," the fact, according to Tallant (1946, 93), is that "despite their frequency in fiction about Voodoo, dolls were rarely used in the practices." Nevertheless, today they are everywhere. One can at least shun the made-in-China souvenirs for the local variety sold at Marie Laveau's House of Voodoo, Rev. Zombie's Voodoo Shop, and the New Orleans Historic Voodoo Museum. The latter attraction is well worth seeing for its display of historic artifacts relating to voodoo and its practitioners, including Marie Laveau.

The museum promotes voodoo—including its commercial, tourist aspect—through various offerings, including annual rituals on St. John's Eve and Halloween, and for-hire performances offered as party entertainment. Walking tours of voodoo-related sites in the Vieux Carré are also available daily.

Tour groups may routinely visit the Voodoo Spiritual Temple on Rampart Street at the edge of the Vieux Carré. I enlisted a professional guide and was able to gain an audience with Priestess Miriam, perhaps today's premier voodoo queen. At the end of an interesting visit, she waived the prohibition against photographs and permitted me to document some of the authentic voodoo altars of this religious and cultural

FIGURE 17-2. **Authentic "working" altar in New Orleans' Voodoo Spiritual Temple. Candles, religious effigies, bottles of rum (as offerings), and other elements are typical.**

center (Figure 17-2). These are "working" altars, meaning that they are used in rituals and are modified to invoke and propitiate various spirits.

Tours also take visitors to the reputed tomb of Marie Laveau, where they may hope to have a wish granted or glimpse her ghost, which allegedly haunts the site. (See also chapter 18, "Secrets of the Voodoo Tomb.") Although voodoo has declined from the early days, when Marie held New Orleans under her spell, her influence nevertheless continues.

NOTE

1. Marie apparently had one child by her first husband and five more by her second (who was reportedly white but "passed" as a free person of color).

The 15 children usually attributed to her is an error; the additional nine were those of Marie's half-sister of the same name (Jensen 2002).

REFERENCES

"America's Haunted Houses." 1998. Discovery Channel. First aired 24 May.

Antippas, A. P. 1988. *A Brief History of Voodoo: Slavery & the Survival of the African Gods*. New Orleans, La.: Marie Laveau's House of Voodoo.

Bourguignon, Erika E. 1993. "Voodoo." In *Collier's Encyclopedia*. New York: P. F. Collier.

"Death of Marie Laveau." 1881. Obituary in *Daily Picayune*, n.d. (after June 15); clipping reproduced in Gandolfo 1992, 38; text quoted in full in Tallant 1946, 113–16.

Gandolfo, Charles. 1992. *Marie Laveau of New Orleans*. New Orleans, La.: New Orleans Historic Voodoo Museum.

———. N.d. Museum guide sheet. New Orleans Historic Voodoo Museum.

Hauck, Dennis William. 1996. *Haunted Places: The National Directory*. New York: Penguin Books.

Hurbon, Laënnec. 1995. *Voodoo: Search for the Spirit*. New York: Harry N. Abrams.

Hyman, Ray. 1977. Cold-reading: How to convince strangers that you know all about them. *Skeptical Inquirer* 1, no. 2 (Spring/Summer), 18–37.

Jensen, Lynne. 2002. LSU professor finds Laveau's birth records. *The Advocate* (Baton Rouge, La.), Februar 25, 5B.

Keene, M. Lamar. [1976] 1997. *The Psychic Mafia*. Reprinted Amherst, N.Y.: Prometheus Books, 1997.

Klein, Victor C. 1999. *New Orleans Ghosts II*. Metairie, La.: Lycanthrope Press.

"Marie Lavaux [sic]." 1881. Obituary in *New Orleans Democrat*, 17 June, reproduced in Gandolfo 1992, 37.

Martinez, Raymond J. [1956] N.d. *Mysterious Marie Laveau, Voodoo Queen, and Folk Tales Along the Mississippi*. Reprinted New Orleans: Hope Publications.

Métraux, Alfred. 1972. *Voodoo in Haiti*. New York: Schocken Books.

Nardo, Don, and Erik Belgum. 1991. *Great Mysteries: Voodoo: Opposing Viewpoints*. San Diego, Cal.: Greenhaven Press.

Tallant, Robert. [1946] 1990. *Voodoo in New Orleans*. Reprinted Gretna, La.: Pelican Publishing.

18

Secrets of the Voodoo Tomb

Among the sites associated with New Orleans voodoo is the tomb of its greatest figure, Marie Laveau (the subject of chapter 17). After the apparent death of her first husband, Jacques Paris, she began calling herself the "Widow Paris." (The relevance of this will soon be apparent.)

The Wishing Tomb

Controversy persists over where Marie Laveau and her namesake daughter are buried. Some say the latter reposes in the cemetery called St. Louis No. 2 (Hauck 1996) in a "Marie Laveau Tomb" there. However, that crypt most likely contains the remains of another voodoo queen named Marie, Marie Comtesse. Numerous sites in as many cemeteries are said to be the final resting place of one or the other Marie Laveau (Tallant 1946, 129), but the *prima facie* evidence favors the Laveau-Glapion tomb in St. Louis No. 1 (Figure 18-1). It comprises three stacked crypts with a "re-

FIGURE 18-1. **Laveau-Glapion tomb in New Orleans' St. Louis Cemetery No. 1. (Photograph by Joe Nickell.)**

ceiving vault" below (that is, a repository of the remains of those displaced by a new burial).

A contemporary of Marie II told Tallant (1946, 126) that he had been present when Marie II died of a heart attack at a ball in 1897, and insisted: "All them other stories ain't true. She was buried in the Basin Street graveyard they call St. Louis No. I, and she was put in the same tomb with her mother and the rest of her family." (Except as otherwise noted, information about Marie Laveau and her daughter is taken from Tallant.)

A carved inscription on the St. Louis No. 1 tomb records the name, date of death, and age (62) of Marie II: "Marie Philome Glapion, décédé le 11 Juin 1897, ágée de Soixante-deux ans." A bronze tablet affixed to the tomb announces, under the heading "Marie Laveau," that "This Greek Revival Tomb Is Reputed Burial Place of This Notorious 'Voodoo Queen'"—presumably a reference to the original Marie. (See Figure 18-2.) Corroborative evidence that she was interred here is found in her obituary ("Death" 1881), which notes that "Marie Laveau was buried in her family tomb in St. Louis Cemetery No. 1." Guiley (2000, 213–16) asserts that although Marie Laveau I is reportedly buried here, "[t]he vault does not bear her name." However, I was struck by the fact that the initial two lines of the inscription on the Laveau-Glapion tomb read, "Famille Vve. Paris / née Laveau." Obviously, "Vve." is an abbreviation for *Veuve*, "Widow"; therefore, the phrase translates, "Family of the Widow Paris, born Laveau"—namely Marie Laveau I. I take this as evidence that here is indeed the "family tomb." Robert Tallant (1946, 127) suggests: "Probably there was once an inscription marking the vault in which the first Marie was buried, but it has been changed for one marking a later burial. The bones of the Widow Paris must lie in the receiving vault below."

The Laveau-Glapion tomb is a focal point for commercial voodoo tours. Some visitors leave small gifts at the site—coins, Mardi Gras beads, candles, and the like—in the tradition of voodoo offerings. Many follow a custom of making a wish at the tomb. The necessary ritual for this has been variously described. The earliest version I have found (Tallant 1946, 127) says that people would "knock three times on the slab and ask a favor," noting: "There are always penciled crosses on the slab. The sexton washes the crosses away, but they always reappear." A more recent source advises combining the ritual with an offering placed in the attached cup: "Draw the X, place your hand over it, rub your foot

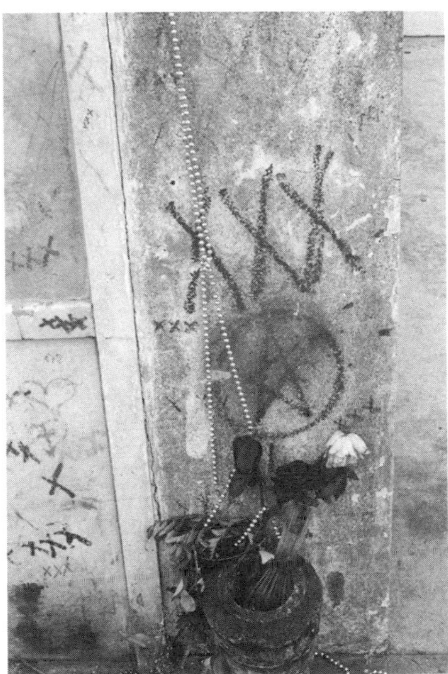

FIGURE 18-2. **Tablet placed on the tomb by the Catholic Archdiocese follows tradition in identifying the site of Marie Laveau's burial.**

FIGURE 18-3. **Sets of three Xs and other markings—together with offerings of flowers, beads, money, and other items—decorate the reputed tomb of the celebrated Voodoo Queen. (Photographs by Joe Nickell.)**

three times against the bottom, throw some silver coins into the cup, and make your wish" (Haskins 1990, 59–61). Yet another source says that petitioners are to "leave offerings of food, money and flowers, then ask for Marie's help after turning around three times and marking a cross with red brick on the stone" (Guiley 2000, 216).

When I visited the tomb, it was littered with markings, including single Xs; an occasional cross, heart, pentagram, or other figure; and a few inscriptions or other graffiti, sometimes accompanied by initials (Figure 18-3). One comment read: "Her eyes / lit up with Fire / For the dreams / she entertained . . . / Seems something in her / knew already /

just how well / They'd burn. / A.R.P. / 11–19–00." The predominant markings were sets of three Xs—suggesting that the folk practice is undergoing transition (that is, the specified number of raps or turns is apparently becoming transferred to the number of Xs to be inscribed).

Although some of the markings are done in black (as from charcoal), most are rendered in a rusty red from bits of crumbling brick. One New Orleans guidebook says of the wishing tomb:

> The family who own it have asked that this bogus, destructive tradition should stop, not least because people are taking chunks of brick from other tombs to make the crosses. Voodoo practitioners—responsible for the candles, plastic flowers, beads, and rum bottles surrounding the plot—deplore the practice, too, regarding it as a desecration that chases Laveau's spirit away [Cook 1999, 110, 112].

Echoing that view, another guidebook advises: "On the St. Louis tour, please don't scratch Xs on the graves; no matter what you've heard, it is not a real voodoo practice and is destroying fragile tombs" (Herczog 2000, 158, 186).

The scuttlebutt, according to the professional guide I commissioned (Krohn 2000), is that the practice may have evolved from the appearance of ordinary graffiti, which was then transformed by an early cemetery guide into a pseudo-voodoo custom that brought him tips. Of the wishing practice, one writer wryly observes that there is "no word on success rates" (Dickinson 1997).

Perturbed Spirit

Given the belief that Marie Laveau's spirit can be invoked to grant wishes, it was inevitable that there would be alleged sightings of her ghost. According to the author of *Haunted City* (Dickinson 1997, 131):

> Tour guides tell of a Depression-era vagrant who fell asleep atop a tomb in the cemetery and was awakened to the sound of drums and chanting. Stumbling upon the tomb of Marie Laveau, he encountered the ghosts of dancing, naked men and women, led by a tall woman wrapped in the coils of a huge snake.

Or so tour guides tell. But did the "vagrant" perhaps pass out from drink and have a vivid dream or hallucination? How much has the story been embellished in the intervening two-thirds century or so? Do we know that the alleged event even occurred? These are among the problems with such anecdotal evidence.

The Encyclopedia of Ghosts and Spirits asserts: "One popular legend holds that Marie I never died, but changed herself into a huge black crow which still flies over the cemetery." Indeed, "[b]oth Maries are said to haunt New Orleans in various human and animal forms" (Guiley 2000). Note the anonymity inherent in such phrases as "popular legend" and the passive-voice construction of "are said to." In addition to her tomb, Marie also allegedly haunts other sites. For example, according to Hauck (1996, 192, 193), "Laveau has also been seen walking down St. Ann Street wearing a long white dress." Providing a touch of what literary critics call verisimilitude (an appearance of truth), Hauck adds, "The phantom is that of the original Marie, because it wears her unique tignon, a seven-knotted handkerchief, around her neck." But Hauck has erred: Marie in fact "wore a large white headwrap called a tignon tied around her head," says her biographer Gandolfo (1992, 19), which had "seven points folded into it to represent a crown." Gandolfo, who is also an artist, has painted a striking portrait of Marie Laveau wearing her tignon, which is displayed in the gift shop of his New Orleans Historic Voodoo Museum (and is reproduced in Gandolfo 1992, 1).

With a bit of literary detective work, we can track the legend-making process in one instance of Laveau ghost lore. In his *Haunted Places: The National Directory*, Hauck (1996, 192, 193) writes of Marie: "Her ghost and those of her followers are said to practice wild voodoo rituals in her old house." Are said to by whom? Hauck's list of sources for the entry on Marie Laveau includes Susy Smith's *Prominent American Ghosts* (1967, 139–40), his earliest-dated citation. Smith merely says of Marie, "Her home at 1020 St. Ann Street was the scene of weird secret rites involving various primitive groups," though she asks, "May not the wild dancing and pagan practices still continue, invisible, but frantic as ever?" Apparently this purely rhetorical question about possible ghosts was transformed into an "are-said-to"-sourced assertion about supposedly real ones. In fact, the house at 1020 St. Ann Street was never even occupied by Marie Laveau; it only marks the approximate site of the home she lived in until her death (then numbered 152 Rue St. Ann, as shown by her death certificate). That cottage, which bore a red-tile roof and was flanked by banana trees and an herb garden, was demolished in 1903 (Gandolfo 1992, 14–15, 34).

Many of the tales of Marie Laveau's ghost, if not actually invented by tour guides, may be uncritically promulgated by them. According to *Frommer's New Orleans 2001*, "We enjoy a good nighttime ghost tour of the Quarter as much as anyone, but we also have to admit that what's available is really hit-or-miss in presentation (it depends on who conducts your particular tour) and more miss than hit with regard to facts" (Herczog 2000, 158, 186). Even the author of *New Orleans Ghosts II*—hardly a knee-jerk debunker—speaks of the "hyperbolic balderdash" that sometimes "spews forth from the black garbed tour guides who are more interested in money and sensationalism than accurate historical research" (Klein 1999, 64).

A Haunting Tale

One alleged Laveau ghost-sighting stands out. Tallant (1946, 130–31) relates the story of an African-American named Elmore Lee Banks, who had an experience near St. Louis Cemetery No. 1. As Banks recalled, one day in the mid-1930s "an old woman" came into the drugstore where he was a customer. For some reason she frightened the proprietor, who "ran like a fool into the back of the store." Laughing, the woman asked, "Don't you know me?" She became angry when Banks replied, "No, ma'am," and slapped him. Banks continued: "Then she jump[ed] up in the air and went whizzing out the door and over the top of the telephone wires. She passed right over the graveyard wall and disappeared. Then I passed out cold." He awakened to whiskey being poured down his throat by the proprietor, who told him, "That was Marie Laveau."

What are we to make of this case? (Perhaps the reader will want to pause here and reflect on the possibilities.) Let us assume, provisionally, that such an event did transpire, although we can predict that the narrative may have been affected by the well-known influences of misperception, memory distortion, the unconscious temptation to embellish, and other factors. We can begin our analysis by noting a few clues. First, it seems significant that Banks was a customer in a drugstore; this suggests that he may have been ill and/or on medication. Second, it seems curious that he "passed out cold" from a mere slap, especially a ghostly one. (It seems contradictory that ghosts—which are reputedly nonphysical and often are reported to be able to pass through walls—are able to perform physical acts.) A third clue, I think, comes from the contrast between the first part of the story, wherein the woman appears quite unghostlike and acts in concert with the real world, and the second part, in which her behavior (flying through the air) seems consistent with an hallucinatory experience.

Putting the clues together gives us the following possible scenario: Banks visits the drugstore because he is unwell, possibly seeking to get a prescription filled or refilled. An elderly woman comes in, recognizes him (perhaps from some years before), and is bemused that he fails to recognize her. Suddenly, from the effects of his illness or medication or even alcohol, Banks passes out, but in the process of swooning and falling to the floor he hallucinates. This may have involved his brain perceiving the lowering of his body in relationship to hers as the converse action—as her rising above him—and so triggering a dream-like fantasy of her flying. (Hallucinations can occur in normal individuals with various medical conditions, including high fevers and reduced respiration rates, as well as alcoholic states and many other conditions. And hallucinations "share much in common with dreams" [Baker 1992, 274–76].)

The various elements in the story may have become confused—misconstrued and misordered as to sequence—as Banks teetered on the brink of consciousness. For example, although the woman may have slapped him in anger, another possibility is that she did so slightly later in an attempt to revive him. Similarly, the proprietor may have run to the rear of the store not because he recognized the "ghost," but in order to fetch the whiskey with which to revive Banks. Subsequently, while seeming to have "witnessed" the entire event (Hauck 1996, 192, 193) and to have identified the woman as Marie Laveau, the store owner may in fact only have been commenting on the perceived events that Banks related. Over time, as Banks repeated and rehearsed his tale, it became a dramatic, supernatural narrative about Marie Laveau. States psychologist Robert A. Baker: "The work of Elizabeth Loftus and others over the past decade has demonstrated that the human memory works not like a tape recorder but more like the village storyteller, i.e., it is both creative and recreative" (Baker and Nickell 1992, 217).

Such impulses may be especially strong in a climate of magical thinking. They have helped foster the many tales and claims about Marie Laveau. In addition, according to the *Encyclopedia of African-American Culture and History* (Salzman 1996, 1581), "the legend of Marie Laveau was kept alive by twentieth-century conjurers who claimed to use Laveau techniques and it is kept alive through the continuing practice of commercialized voodoo in New Orleans."

REFERENCES

Baker, Robert A. 1992. *Hidden Memories: Voices and Visions from Within.* Buffalo, N.Y.: Prometheus Books.

Baker, Robert A., and Joe Nickell. 1992. *Missing Pieces: How to Investigate Ghosts, UFOs, Psychics, and Other Mysteries.* Buffalo, N.Y.: Prometheus Books.

Cook, Samantha. 1999. *New Orleans: The Mini Rough Guide.* London: Rough Guides Ltd.

"Death of Marie Laveau." 1881. Obituary, *Daily Picayune*(New Orleans, La.), n.d. (after June 15), reprinted in Gandolfo 1992, 38–39.

Dickinson, Joy. 1997. *Haunted City: An Unauthorized Guide to the Magical, Magnificent New Orleans of Anne Rice.* Secaucus, N.J.: Citadel Press.

Gandolfo, Charles. 1992. *Marie Laveau of New Orleans.* New Orleans, La.: New Orleans Historic Voodoo Museum.

Guiley, Rosemary Ellen. 2000. *The Encyclopedia of Ghosts and Spirits.* 2d ed. New York: Checkmark Books.

Haskins, Jim. 1990. *Voodoo & Hoodoo.* New York: Scarborough House.

Hauck, Dennis William. 1996. *Haunted Places: The National Directory.* New York: Penguin Books.

Herczog, Mary. 2000. *Frommer's 2001 New Orleans.* New York: IDG Books Worldwide.

Klein, Victor. 1999. *New Orleans Ghosts II.* Metairie, La.: Lycanthrope Press.

Krohn, Diane C. 2000. Personal communication, 3 December.

Salzman, Jack, et al., eds. 1996. *Encyclopedia of African-American Culture and History*, vol. 3. London: Simon & Schuster and Prentice Hall International.

Smith, Susy. 1967. *Prominent American Ghosts.* Cleveland, Ohio: World Publishing.

Tallant, Robert. [1946] 1990. *Voodoo in New Orleans*, reprinted Gretna, La.: Pelican Publishing.

19

A Case of "SHC" Demystified

A case of March 4, 1980, in Chorley, England, mystifies paranormalists, who invoke spontaneous human combustion (SHC) in a fatal accident. Where is the mystery? Tony McMunn, a fireman who encountered the case and became an SHC enthusiast as a result, insists that "there is not a lot of flesh or fat on the head, and the fire should have gone out." He and others are also impressed by the severe destruction of the body, in which some of the bones were reportedly calcined (reduced to ash). However, the following investigative chronology, keyed to the pen-and-ink drawing shown in Figure 19-1 and based on a published photograph, easily resolved the mystery. (See Randles and Hough 1992, 84–85, 91, and illus. 6 following p. 112.)

1. Bucket indicated to investigators that the victim, an elderly lady, was in the process of relieving herself when she fell.

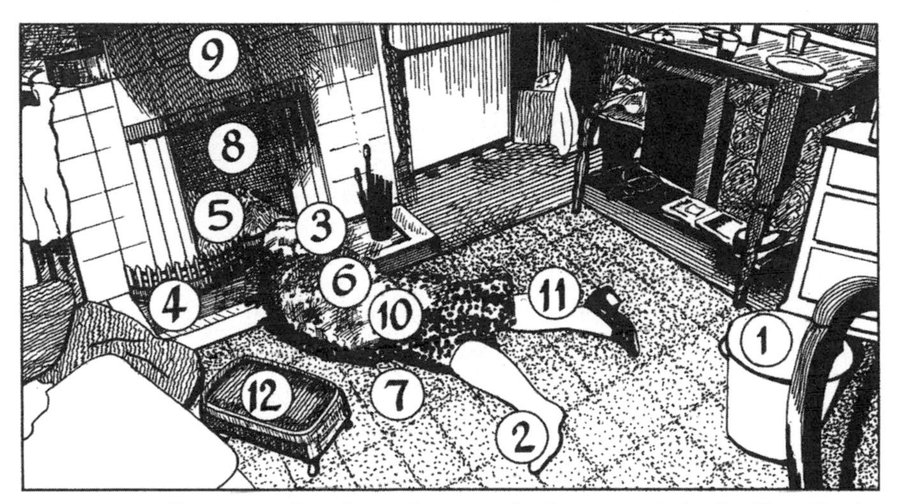

Figure 19-1. Illustration based on photograph of a woman's body allegedly destroyed by spontaneous human combustion.

2. The missing shoe is consistent with this or other possible scenarios. Apparently it came off during the victim's fall and is out of view; it may be that her attempts to take it off even *caused* the fall.
3. In falling, the victim obviously hit her head on the fireplace; the impact either knocked her unconscious or possibly killed her outright.
4. Her head struck the iron grate, which has been sharply displaced to the left.
5. The fall caused flaming embers from the now-exposed "open coal fire" to shower upon the body.
6. The victim's clothing ignited. As the fire progressed, her own melting body fat contributed to the overall destruction.
7. The rug beneath the body may have retained melted body fat and thus aided in the severe destruction—a process known in the forensic literature as the *wick effect*.
8. The fire was probably further enhanced by the *chimney effect*—a "drawing" of the flame and venting of smoke; in this case, the

venting took place through the chimney itself. At about 9:30 on the preceding evening, when the fire was believed to have taken place, neighbors saw a great amount of smoke and sparks issuing from the chimney.

9. Heavy deposits of soot above the fireplace, tapering toward the chimney opening, are consistent with the chimney effect and the venting of considerable organic material.
10. The destruction of the body was in approximate proportion to its proximity to the fire source. The torso—which contains a large amount of fat—was most severely destroyed, whereas the lower legs and feet remained intact.
11. As in many other such cases, the lower extremities were spared because fire burns laterally only with difficulty.
12. Nearby objects failed to burn for the same reason. Only radiant heat, not flame, reached these objects.

The only mystery remaining is why anyone would ever have considered that the death occurred by spontaneous human combustion!

REFERENCES

Randles, Jenny, and Peter Hough. 1992. *Spontaneous Human Combustion*. London: Robert Hale.

20

Tracking the Swamp Monsters

Do mysterious and presumably endangered manlike creatures inhabit swamplands of the southern United States? If not, how do we explain the sightings and even track impressions of creatures that thus far have eluded mainstream science? Do they represent additional evidence of the legendary Bigfoot, or something else entirely? What would an investigation reveal?

Monster Mania

The outside world learned about Louisiana's Honey Island Swamp Monster in 1974, when two hunters emerged from a remote area of backwater sloughs with plaster casts of "unusual tracks." The men claimed that they had discovered the footprints near a wild boar that lay with its throat gashed. They also stated that more than a decade earlier, in 1963, they had seen similar tracks after encountering an awesome creature. They described it as standing seven feet tall, being covered with grayish hair, and having large amber-colored eyes. However, the monster had promptly run away and an afternoon rainstorm had obliterated its tracks, the men said.

The hunters were Harlan E. Ford and his friend Billy Mills, both of whom worked as air traffic controllers. Ford told his story on an episode of the 1970s television series *In Search of*. According to his granddaughter, Dana Holyfield (1999a, 11):

> When the documentary was first televised, it was monster mania around here. People called from everywhere. . . . The legend of the Honey Island Swamp Monster escalated across Southern Louisiana and quickly made its way out of state after the documentary aired nationwide.

Harlan Ford continued to search for the monster until his death in 1980. Dana recalls how he once took a goat into the swamp to use as bait, hoping to lure the creature to a tree blind where Ford waited with gun and camera—uneventfully, as it happened. He supposedly did find several different-sized tracks on one hunting trip. He also claimed to have seen the monster on one other occasion, during a fishing trip with Mills and some of their friends from work. One of the men reportedly then grabbed a rifle and went searching for the creature, and fired two shots at it before returning to tell his story to the others around the campfire (Holyfield 1999a, 10–15).

Searching for Evidence

Intrigued by the monster reports, which I pursued on a trip to New Orleans (to speak to local skeptics at the planetarium in Kenner), I determined to visit the alleged creature's habitat. The Honey Island Swamp (Figure 20-1) comprises nearly 70,000 acres between the East Pearl and West Pearl Rivers. I signed on with Honey Island Swamp Tours, which

FIGURE 20-1. **Louisiana's pristine Honey Island Swamp is the alleged habitat of a manlike monster. (Photograph by Joe Nickell.)**

is operated out of Slidell, Louisiana, by wetlands ecologist Dr. Paul Wagner and his wife, Sue. Their "small, personalized nature tours" live up to their billing as explorations of "the deeper, harder-to-reach small bayous and sloughs" of "one of the wildest and most pristine river swamps in America" ("Dr. Wagner's" n.d.).

The Wagners are ambivalent about the existence of the supposed swamp monster. They have seen alligators, deer, otters, bobcats, and numerous other species, but not a trace of the legendary creature (Wagner 2000). The same is true of the Wagners' Cajun guide, Captain Robbie Charbonnet. Beginning at age 8, he has had 45 years' experience (18 as a guide) in the Honey Island Swamp. He told me he had "never seen or heard" something he could not identify, certainly nothing that could be attributed to a monster (Charbonnet 2000).

Suiting action to words throughout our tour, Charbonnet repeatedly identified species after species in the remote swampland, as he skillfully threaded his boat through the cypresses and tupelos hung with Spanish moss. Although the cool weather had pushed alligators to the depths, he heralded turtles, great blue herons, and other wildlife. From only a glimpse of its silhouetted form, he spotted a barred owl, then carefully maneuvered for a closer view. He called attention to the singing of robins, which were gathering in the swamp for the winter, and pointed to signs of other creatures, including branches freshly cut by beavers and tracks left in the mud by a wild boar. But there was not a trace of any swamp monster. (The closest I came was passing an idle boat at Indian Village Landing emblazoned "Swamp Monster Tours.")

Another who is skeptical of monster claims is naturalist John V. Dennis. In his comprehensive book, *The Great Cypress Swamps* (1988, 27, 108–9), he writes: "Honey Island has achieved fame of sorts because of the real or imagined presence of a creature that fits the description of the Big Foot of movie renown. Known as the Thing, the creature is sometimes seen by fishermen." However, he says, "For my part, let me say that in my many years of visiting swamps, many of them as wild or wilder than Honey Island, I have never obtained a glimpse of anything vaguely resembling Big Foot, nor have I ever seen suspicious-looking footprints." He concludes, "Honey Island, in my experience, does not live up to its reputation as a scary place."

In contrast to the swamp experts' lack of monster experiences are the encounters reported by Harlan Ford and Billy Mills. Those alleged eyewitnesses are, in investigators' parlance, "repeaters"—people who claim unusual experiences on multiple occasions. (Take Bigfoot hunter Roger Patterson, for example. Before shooting his controversial film sequence of a hairy man-beast in 1967, Patterson was a longtime Bigfoot buff who had "discovered" the alleged creature's tracks on several occasions [Bord and Bord 1982, 80].) Ford's and Mills's multiple sightings and discoveries seem suspiciously lucky, and suspicions are increased by other evidence, including the tracks.

From Dana Holyfield I obtained a plaster copy of one of the several

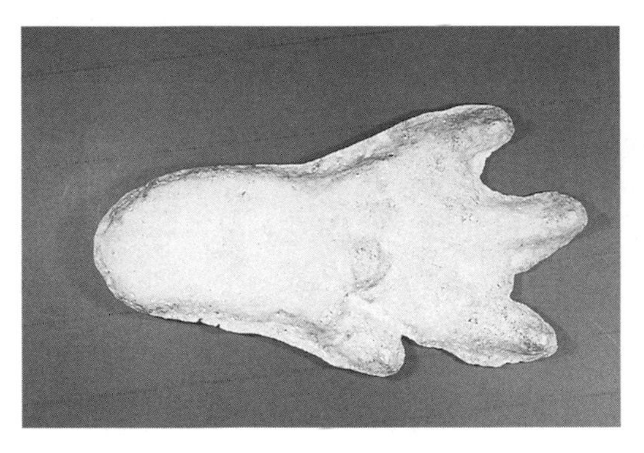

FIGURE 20-2. **This plaster cast preserves an alleged Honey Island Swamp Monster track.**

track casts made by her grandfather (Figure 20-2). It is clearly not the track of a stereotypical Bigfoot (or sasquatch), whose footprints are "roughly human in design," according to anthropologist and pro-Bigfoot theorist Grover Krantz (1992, 17). Instead, Ford's monster tracks are webbed-toe imprints that appear to be "a cross between a primate and a large alligator" (Holyfield 1999a, 9). The track is also surprisingly small: only about 9.75 inches long. Compare this to alleged Bigfoot tracks, which average about 14 to 16 inches (Coleman and Clark 1999, 14), with tracks of 20 inches and more reported (Coleman and Huyghe 1999, 14–19).

Monsterlands

Clearly, the Honey Island Swamp Monster is not a Bigfoot, a fact that robs Ford's and Mills's story of any credibility it might have gained from that association. Monster popularizers instead equate the Honey Island reports with other "North American 'Creatures of the Black Lagoon' cases" in which there is purported evidence of cryptozoological entities, dubbed "freshwater Merbeings" (Coleman and Huyghe 1999, 39, 62). These cases are supposedly linked by tracks with three toes, although Ford's casts actually exhibit four (again see Figure 20-2). In short, this alleged monster is unique, rare even among creatures whose existence is unproven and unlikely.

Footprints and other specific details aside, the Honey Island Swamp Monster seems part of a genre of mythic swamp-dwelling "beastmen" or "manimals." They include the smelly Skunk Ape and the hybrid Gatorman of the Florida Everglades and other southern swamps; the Scape Ore Swamp Lizardman of South Carolina; Momo, the Missouri Monster; and, among others, the Fouke Monster, which peeked in the window of a home in Fouke, Arkansas, one night in 1971 and set off a rash of monster sightings (Blackman 1998, 23–25, 30–33, 166–68; Bord and Bord 1982, 104–5; Coleman and Clark 1999, 224–26; Coleman and Huyghe 1999, 39, 56).

Considering this genre, we must ask: Why swamps and why monsters? Swamps represent remote, unexplored regions, which traditionally are the domain of legendary creatures. As the noted Smithsonian Institution biologist John Napier (1973, 23) sagely observed, monsters "hail from uncharted territory: inaccessible mountains, impenetrable forests, remote Pacific islands, the depths of loch or ocean. . . . The essential element of the monster myth is remoteness."

Echoing Napier in discussing one reported Honey Island Swamp encounter, John V. Dennis (1988) states: "In many cases, sightings such as this one are inspired by traditions that go back as far as Indian days. If a region is wild and inaccessible and has a history of encounters with strange forms of life, chances are that similar encounters will occur again—or at least be reported." Although the major purported domain of Bigfoot is the Pacific Northwest, Krantz (1992, 199) observes: "Many of the more persistent eastern reports come from low-lying and/or swampy lands of the lower Mississippi and other major river basins."

But why does belief in monsters persist? According to one source, monsters appear in every culture and are "born out of the unknown and nurtured by the unexplained" (Guenette and Guenette 1975). Many alleged paranormal entities appear to stem either from mankind's hopes or fears, and thus are envisioned as angels and demons; some entities may evoke a range of responses. Monsters, for example, may both intrigue us with their unknown, mysterious aspects and provoke terror. We may be especially interested in man-beasts, given what psychologist Robert A. Baker (1995) observes is our strong tendency to impute human characteristics to nonhuman things and entities. Hence, angels are basically our better selves with wings; extraterrestrials are humanoids from futuristic worlds; and Bigfoot and his ilk seem linked to our evolutionary past.

Monsters may play various roles in our lives. My Cajun guide, Robbie Charbonnet, offered some interesting ideas about the Honey Island Swamp Monster and similar entities. He thought that frightening stories might have been concocted on occasion to keep outsiders away—perhaps to protect prime hunting areas or even help safeguard moonshine stills. He also theorized that such tales might have served, in a sort of bogeyman fashion, to frighten children enough to keep them from wandering into remote, dangerous areas. (Indeed, he mentioned that when he was a youngster in the 1950s, an uncle would tell him about a frightening figure—a sort of horror-movie type with one leg, a mutilated face, and so on—that would "get" him if he strayed into the swampy wilderness.)

Like any such bogeyman, the Honey Island Swamp Monster is also good for gratuitous campfire chills. "A group of men were sitting around the campfire along the edge of the Pearl River," begins one narrative, "telling stories about that thing in the swamp . . ." (Holyfield 1999b). A song, "The Honey Island Swamp Monster" (written by Perry Ford and quoted in Holyfield 1999b, 13), is in a similar vein: "Late at night by a dim fire light, / You people best beware. / He's standing in the shadows, / Lurking around out there . . . ." The monster has even been referred to specifically as "The boogie man" and "that booger" (Holyfield 1992a, 14). "Booger" is a dialect form of *bogey*, and deliberately scary stories are sometimes known as "'booger' tales" (Cassidy 1985, 333–34).

Suitable subjects for booger tales are the numerous Louisiana swamp and bayou terrors, many of which are the products of Cajun folklore. One is the Letiche, a ghoulish creature that was supposedly an abandoned, illegitimate child who was reared by alligators, and now has scaly skin, webbed hands and feet, and luminous green eyes. Then there is Jack O'Lantern, a malevolent spirit who lures humans into dangerous swampland with his mesmerizing lantern, as well as the loup garou (a werewolf) and zombies (not the relatively harmless "Voodoo Zombies" but the horrific "Flesh Eaters") (Blackman 1998, 171–209).

By extension, swamp creatures are also ideal subjects for horror fiction. The Fouke monster sightings, for example, inspired the horror movie *The Legend of Boggy Creek*. That 1972 thriller became a box-office hit, spawning a sequel and many imitations. At about the same time (1972), there emerged a popular comic book series titled *Swamp Thing*, featuring a metamorphosing man-monster from a Louisiana swamp. Interestingly, these popularized monsters predated the 1974 claims of Ford and Mills. (Recall that their alleged earlier encounter of 1963 had not been reported until after the second encounter in 1974.)

The Track Makers

Although swamp monsters and other man-beasts have not been proven to exist, hoaxers certainly have. Take, for example, the Bigfoot tracks reported by berry pickers near Mount St. Helens, Washington, in 1930. Nearly half a century later, a retired logger came forward to pose with a set of "bigfeet" that he had carved and that a friend had worn to produce the fake monster tracks (Dennett 1982). Among many similar hoaxes were at least seven perpetrated in the early 1970s by one Ray Pickens of Chehalis, Washington. He carved primitive seven-by-eighteen-inch feet and attached them to hiking boots. Pickens (1975) said he was motivated "not to fool the scientists, but to fool the monster-hunters," whom he felt regarded people like him as "hicks" (Guenette and Guenette 1975, 80). Other motivation, according to monster hunter Peter

Byrne, stems from the "extraordinary psychology of people wanting to get their names in the paper, people wanting a little publicity, wanting to be noticed." (Guenette and Guenette 1975, 81).

Were Harlan Ford's and Billy Mills's monster claims similarly motivated? Dana Holyfield (1999a, 5–6) says of her grandfather: "Harlan wasn't a man to make up something like that. He was down to earth and honest and told it the way it was and didn't care if people believed him or not." But even a basically honest person, who would not do something overtly dastardly or criminal, might engage in a prank or hoax that he considered relatively harmless and that would add zest to life. I believe the evidence strongly indicates that Ford and Mills did just that. To sum up, there are the men's suspiciously repeated reports or sightings and their alleged track discoveries, together with the incongruent mixing of a Bigfoot-type creature with most un-Bigfootlike feet, plus the fact that the proffered evidence is of a type that can easily be faked and often has been. In addition, the men's claims exist in a context of swamp-manimal mythology that has numerous antecedent elements in folklore and fiction. Taken together, the evidence suggests a common hoax.

Certainly, in the wake of the monster mania Ford helped inspire, much hoaxing resulted. States Holyfield (1999a, 11):

> Then there were the monster impersonators who made fake bigfoot shoes and tromped through the swamp. This went on for years. Harlan didn't worry about the jokers because he knew the difference.

Be that as it may, swamp-monster hoaxes—and apparent hoaxes—continue.

A few months before I arrived in Louisiana, two loggers, Earl Whitstine and Carl Dubois, reported sighting a hairy man-beast in a cypress swamp called Boggy Bayou in the central part of the state. Giant four-toed tracks and hair samples were discovered at the site, and soon others came forward to say that they too had seen a similar creature. However, there were grounds for suspicion: 25 years earlier (i.e., not long after the 1974 Honey Island Swamp Monster reports), Whitstine's father and some friends had sawed giant foot shapes from plywood and produced fake monster tracks in the woods of a nearby parish.

On September 13, 2000, laboratory tests of the hair from the Boggy Bayou creature revealed that it was not from a *Gigantopithecus blacki* (a scientific name for the sasquatch proposed by Krantz [1992, 193]), but much closer to *Booger louisiani* (my term for the legendary swamp bogeyman). It proved actually to be from *Equus caballus* (a horse)—whereupon the local sheriff's department promptly ended its investigation (Blanchard 2000, Burdeau 2000).

Reportedly, Harlan Ford believed the swamp monsters "were probably on the verge of extinction" (Holyfield 1999a, 10). Certainly he did much to further their cause. It seems likely that as long as there are suitably remote habitats and other essentials (such as campfires around which to tell tales, and good ol' boys looking for their 15 minutes of fame), the legendary creatures will continue to proliferate.

NOTE

1. Although Harlan Ford obtained tracks of various sizes, a photo of his mounted casts (Holyfield 1999a, 10) makes it possible to compare them with his open hand, which touches the display and thus gives an approximate scale. This shows that all are relatively small. The one I obtained from Holyfield is consistent with the larger ones.

REFERENCES

Baker, Robert A. 1995. Afterword. In *Entities: Angels, Spirits, Demons, and Other Alien Beings,* by Joe Nickell, 275–85. Amherst, N.Y.: Prometheus Books.

Blackman, W. Haden. 1998. *The Field Guide to North American Monsters.* New York: Three Rivers Press.

Blanchard, Kevin. 2000. Bigfoot sighting in La.? *The Advocate* (Baton Rouge, La.), 29 August.

Bord, Janet, and Colin Bord. 1982. *The Bigfoot Casebook*. Harrisburg, Pa.: Stackpole Books.

Burdeau, Cain. 2000. Many in central La. fear Bigfoot. *The Advocate* (Baton Rouge, La.), 15 September.

Cassidy, Frederick G., ed. 1985. *Dictionary of American Regional English*, vol. 1. Cambridge, Mass.: Belknap Press.

Charbonnet, Robbie. 2000. Interview by Joe Nickell, 4 December.

Coleman, Loren, and Jerome Clark. 1999. *Cryptozoology A to Z*. New York: Fireside (Simon & Schuster).

Coleman, Loren, and Patrick Huyghe. 1999. *The Field Guide to Bigfoot, Yeti, and Other Mystery Primates Worldwide*. New York: Avon.

Dennett, Michael. 1982. Bigfoot jokester reveals punchline—finally. *Skeptical Inquirer* 7, no. 1 (Fall): 8–9.

Dennis, John V. 1988. *The Great Cypress Swamps*. Baton Rouge: Louisiana State University Press.

"Dr. Wagner's Honey Island Swamp Tours, Inc." N.d. Advertising flyer. Slidell, La.

Guenette, Robert, and Frances Guenette. 1975. *The Mysterious Monsters*. Los Angeles, Cal.: Sun Classic Pictures.

Holyfield, Dana. 1999a. *Encounters with the Honey Island Swamp Monster*. Pearl River, La.: Honey Island Swamp Books.

———. 1999b. *More Swamp Cookin' with the River People*. Pearl River, La.: Honey Island Swamp Books.

Krantz, Grover. 1992. *Big Footprints: A Scientific Inquiry into the Reality of Sasquatch*. Boulder, Colo.: Johnson Books.

Napier, John. 1973. *Bigfoot: The Yeti and Sasquatch in Myth and Reality*. New York: E.P. Dutton & Co.

Wagner, Sue. 2000. Interview by Joe Nickell, 4 December.

21

John Edward

Talking to the Dead?

Superstar "psychic medium" John Edward is a stand-up guy. Unlike the spiritualists of yore, who typically plied their trade in dark-room séances, Edward and his ilk often perform before live audiences and even under the glare of television lights. Indeed, Edward has his own popular show on the SciFi channel, called *Crossing Over*. I was asked by *Dateline NBC* to study Edward's act and find out if he was really talking to the dead.

The Old Spiritualism

Today's spiritualism traces its roots to the mid-nineteenth century, when the craze spread across the United States, Europe, and beyond. In darkened séance rooms, lecture halls, and theaters, various "spirit" phenomena occurred. For example, the Davenport Brothers conjured up spirit entities to play musical instruments while the two mediums were, apparently, securely tied in a special "spirit cabinet." Unfortunately, the Davenports were exposed many times, once by a local printer. He visited their spook show and volunteered as part of an audience committee to help secure the two mediums. He took that opportunity to surreptitiously place some printer's ink on the a violin; after the séance, one of the spiritualist duo was besmeared with the black substance (Nickell 1999).

The great magician Harry Houdini (1874–1926) crusaded against phony spiritualists, seeking out elderly mediums who taught him the tricks of the trade. For example, although sitters touched hands around the séance table, mediums had clever ways of regaining the use of one hand. (One method was to slowly move the hands close together so that the fingers of one hand could be substituted for those of the other.) This allowed the production of special effects, such as causing a tin trumpet to appear to be levitating. Houdini gave public demonstrations of the deceptions. "Do Spirits Return?" asked one of his posters. "Houdini Says No—and Proves It" (Gibson 1977, 157).

Continuing in the tradition of Houdini, I have investigated various mediums, sometimes attending séances undercover and once obtaining police warrants against a fraudulent medium from the notorious Camp Chesterfield spiritualist center (Nickell 1998). (For more on Camp Chesterfield, see chapter 5, "Undercover Among the Spirits.")

Mental Mediumship

The new breed of spiritualists—like Edward, James Van Praagh, Rosemary Altea, Sylvia Browne, and George Anderson—avoid the physical approach with its risks of exposure and possible criminal charges. They opt instead for the comparatively safe "mental mediumship," which involves the purported use of psychic ability to obtain messages from the spirit realm.

This is not a new approach; mediums have long done "readings" for their credulous clients. In the early days they exhibited "the classic form of trance mediumship, as practiced by shamans and oracles," giving spoken "'spirit messages' that ranged all the way from personal (and sometimes strikingly accurate) trivia to hours-long public trance-lectures on subjects of the deepest philosophical and religious import" (McHargue 1972, 44–45).

Some mediums produced "automatic" or "trance" or "spirit" writing, which the entities supposedly dictated to the medium or produced by guiding his or her hand. Such writings could be in flowery language indeed, as in this excerpt from one spirit writing in my collection:

> Oh my Brother—I am so glad to be able to come here with you and hold sweet communion for it has been a long time since I have controlled this medium but I remember how well used I had become to her magnetism[,] but we will soon get accustomed to her again and then renew the pleasant times we used to have. I want to assure you that we are all here with you this afternoon[—]Father[,] Mother[,] little Alice[—]and so glad to find it so well with you and we hope and feel dear Brother that you have seen the darkest part of life and that times are not with you now as they have been . . . .

and so on in this talkative fashion.

"Cold Reading"

By contrast, today's spirits—whom John Edward and his fellow mediums supposedly contact—seem to have poor memories and difficulty communicating. For example, in one of his on-air séances (on *Larry King Live*, 19 June 1998), Edward said: "I feel like there's a J- or G-sounding name attached to this." He also perceived "Linda or Lindy or Leslie; who's this L name?" Again, he got a "Maggie or Margie, or some M-G-sounding name," and yet again heard from "either Ellen or Helen, or Eleanore—it's like an Ellen-sounding name." Gone is the clear-speaking eloquence of yore; the dead now seem to mumble.

The spirits also seemingly communicate to Edward as if they were engaging in pantomime. As Edward said of one alleged spirit communicant, in a *Dateline* session: "He's pointing to his head; something had to affect the mind or the head, from what he's showing me." No longer, apparently, can the dead speak in flowing Victorian sentences; instead, they are reduced to gestures, as if playing a game of charades.

One suspects, of course, that it is not the imagined spirits that have changed, but rather the approach today's mediums have chosen to employ. It is, indeed, a shrewd technique known as "cold reading"—so named because the subject walks in "cold"; that is, the medium lacks advance information about the person (Gresham 1953, 113–36). It is an artful method of gleaning information from the sitter, then feeding it back as mystical revelation.

The "psychic" can obtain clues by observing dress and body language (noting expressions that indicate when the medium is on or off track), asking questions (which if correct will appear as "hits" but otherwise will seem innocent queries), and inviting the subject to interpret the vague statements offered. For example, nearly anyone can respond to the mention of a common object (like a ring or watch) with a personal recollection that can seem to transform the mention into a hit. (For more on cold reading see Gresham 1953; Hyman 1977; Nickell 2000.)

It should not be surprising that Edward is skilled at cold reading, an old fortunetelling technique. His mother was a "psychic junkie" who threw fortunetelling "house parties." At one of them, an alleged clairvoyant advised the then 15-year-old that he had "wonderful psychic abilities." He began doing card readings for friends and family, then progressed to participating in psychic fairs, where he soon learned that names and other "validating information" sometimes applied to the dead rather than the living. Eventually he changed his billing from "psychic" to "psychic medium" (Edward 1999). The revised approach set him on the road to stardom.

"Hot Reading"

Although cold reading is the main technique of the new spiritualists, they also employ "hot" reading on occasion. Houdini (1924) exposed many of their information-gathering techniques, including the use of planted microphones to listen in on clients as they waited in the mediums' anterooms—a technique Houdini himself used to impress visitors with his "telepathy" (Gibson 1976, 13). Reformed medium M. Lamar Keene's *The Psychic Mafia* (1976) describes such methods as conducting advance research on clients, sharing other mediums' files (what Keene terms "mediumistic espionage"), noting casual remarks made in conversation before a reading, and so on.

An article in *Time* magazine suggested that John Edward may have used just such chicanery. One subject, a marketing manager named Michael O'Neill, had received apparent messages from his dead grandfather—but when his segment aired, he found that it had been improved through editing. According to *Time*'s Leon Jaroff (2001):

> Now suspicious, O'Neill recalled that while the audience was waiting to be seated, Edward's aides were scurrying about, striking up conversations and getting people to fill out cards with their name, family tree and other facts. Once inside the auditorium, where each family was directed to preassigned seats, more than an hour passed before show time while "technical difficulties" backstage were corrected.

Edward has a policy of not responding to criticism, but the executive producer of *Crossing Over* insists: "No information is given to John Edward about the members of the audience with whom he talks. There is no eavesdropping on gallery conversations, and there are no 'tricks' to feed information to John." He labeled the *Time* article "a mix of erroneous observations and baseless theories" (Nordlander 2001).

Be that as it may, on *Dateline* Edward was actually caught in an attempt to pass off previously gained knowledge as spirit revelation. During the session he said of the spirits, "They're telling me to acknowledge Anthony." When the cameraman signaled that that was his name, Edward seemed surprised, asking "That's you? Really?" He further queried: "Had you not seen Dad before he passed? Had you either been away or been distanced?" Later, while playing the taped segment for me, *Dateline* reporter John Hockenberry challenged me with Edward's apparent hit: "He got Anthony. That's pretty good." I agreed but added, "We've seen mediums who mill about before sessions and greet people and chat with them and pick up things."

Indeed, it turned out that that is just what Edward did. Hours before the group reading, Tony had been the cameraman on another Edward shoot (recording him at his hobby, ballroom dancing). Significantly, the two men had chatted, and during the conversation Edward obtained useful bits of information that he afterward pretended had come from the spirits. In a follow-up interview, Hockenberry revealed the fact and grilled an evasive Edward:

Hockenberry: So were you aware that his dad had died before you did his reading?

Edward: I think he—I think earlier in the—in the day, he had said something.

Hockenberry: It makes me feel like, you know, that that's fairly significant. I mean, you knew that he had a dead relative and you knew it was the dad.

Edward: OK.

Hockenberry: So that's not some energy coming through, that's something you knew going in. You knew his name was Tony and you knew that his dad had died and you knew that he was in the room, right? That gets you . . .

Edward: That's a whole lot of thinking you got me doing, then. Like I said, I react to what's coming through, what

I see, hear and feel. I interpret what I'm seeing hearing and feeling, and I define it. He raised his hand, it made sense for him. Great.

Hockenberry: But a cynic would look at that and go, "Hey," you know, "He knows it's the cameraman, he knows it's *Dateline*. You know, wouldn't that be impressive if he can get the cameraman to cry?"

Edward: Absolutely not. Absolutely not. Not at all.

Despite his attempts to weasel out of it, Edward had obviously been caught cheating, pretending that information he had gleaned earlier had just been revealed by spirits and feigning surprise that it applied to Tony the cameraman. (This occurred long before *Time* suggested that an *Inside Edition* program—of 27 February 2001—was probably "the first nationally televised show to take a look at the Edward phenomenon." That honor goes instead to *Dateline NBC*.)

Inflating "Hits"

In addition to shrewd cold reading and out-and-out cheating, "psychics" and "mediums" can also boost their apparent accuracy in other ways. They get something of a free ride from the tendency of credulous folk to count the apparent hits and ignore the misses. In the case of Edward, my analysis of 125 statements or pseudostatements (i.e. questions) he made on a *Larry King Live* program (19 June 1998) showed that he was incorrect about as often as he was right and that his hits were mostly weak ones. For example, he mentioned "an older female" with "an M-sounding name," either an aunt or grandmother, he stated; the caller supplied "Mavis" without identifying the relationship. (Nickell 1998).

Another session—for an episode of *Crossing Over* attended by a reporter for *The New York Times Magazine*, Chris Ballard—had Edward "hitting well below 50 percent for the day." Indeed, he twice spent "upward of 20 minutes stuck on one person, shooting blanks but not accepting the negative responses" (Ballard 2001). This is a common technique: persisting in an attempt to redeem error, cajoling or even browbeating a sitter (as Sylvia Browne often does), or at least making the communication failure seem to be the sitter's fault. "Do not *not* honor him!" Edward exclaimed at one point, then (according to Ballard) "staring down the bewildered man."

When the taped episode actually aired, the two lengthy failed readings had been edited out, along with second-rate offerings. What remained were two of the best readings of the show (Ballard 2001). This seems to confirm the allegation in the *Time* article that episodes were edited to make Edward seem more accurate, even to the point of apparently splicing in clips of one sitter nodding yes "after statements with which he remembers disagreeing" (Jaroff 2001).

Edited or not, group sessions offer increased chances for success. By tossing out a statement and indicating a section of the audience rather than an individual, the performing "medium" makes it many times more likely that someone will "acknowledge" it as a "hit." Sometimes multiple audience members will acknowledge an offering, whereupon the performer typically narrows the choice down to a single person and builds on the success. Edward uses just such a technique (Ballard 2001).

Still another ploy used by Edward and his fellow "psychic mediums" is to suggest that people who cannot acknowledge a hit may find a connection later. "Write this down," an insistent Edward sometimes says, or in some other way suggests that the person study the apparent miss. He may become even more insistent, with the positive reinforcement diverting attention from the failure and giving the person an opportunity to find some adaptable meaning later (Nickell 1998).

Debunking vs. Investigation

Some skeptics believe that the best way to counter Edward and his ilk is to reproduce his effect, by demonstrating the cold-reading technique to radio and television audiences. Of course, that approach is unconvincing unless one actually poses as a medium and then—after seemingly making contact with subjects' dead loved ones—reveals the deception. I deliberately avoid this approach for a variety of reasons, largely because of ethical concerns. I rather agree with Houdini (1924, xi) who did spiritualistic stunts during his early career:

> At the time I did such stunts, I appreciated the fact that I had surprised my clients, but though aware of the fact that I was *deceiving* them, I did not see or understand the seriousness of trifling with such sacred sentimentality—or the unfortunate results that inevitably followed. To me it was a lark. I was a mystifier, my ambition was being gratified, and my love of a mild sensation was being satisfied. After delving deep, though, I realized the seriousness of it all. As I advanced to riper years of experience, I came to realize the seriousness of trifling with the hallowed reverence the average human being bestows on the departed. When I personally became afflicted with similar grief, I was chagrined that I had ever been guilty of such frivolity; for the first time, I realized that it borders on crime.

Of course, tricking people in order to educate them is not the same as deceiving them for crass personal gain, but to toy with their deepest emotions—however briefly and with the best of intentions—is to cross a line that I prefer not to approach. Besides, I believe it can be very counterproductive. It may not be the alleged medium but rather the debunker himself who is perceived as dishonest, and he may come across as arrogant, cynical, and manipulative—not heroic as he imagines.

Furthermore, an apparent reproduction of an *effect* does not necessarily mean that the *cause* was the same. (For example, I have seen several skeptical demonstrations of "weeping" icons that employed trickery more sophisticated than that used for "real" crying effigies.) Far better, I am convinced, is showing evidence of the actual methods employed, as I did in collaboration with *Dateline NBC*.

Although John Edward was among five "highly skilled mediums" who allegedly fared well on a "scientific" test of their ability (Schwartz et al. 2001)—a test critiqued by others (Wiseman and O'Keeffe 2001)—he did not claim validation on *Larry King Live*. When King (2001) asked Edward if he thought there would ever be proof of spirit contact, Edward responded by suggesting that proof is unattainable and that only belief matters: "I think that to prove it is a personal thing. It is like saying, prove God. If you have a belief system and you have faith, then there is nothing really more than that." This, however, is nothing more than an attempt to insulate a position and to evade or shift the burden of proof, which is always on the claimant. As Houdini (1924, 270) emphatically stated, "It is not for us to prove the mediums are dishonest, it is for them to prove that they *are* honest." In my opinion, John Edward has already failed that test.

REFERENCES

Ballard, Chris. 2001. Oprah of the other side. *The New York Times Magazine*, 29 July, 38–41.

Edward, John. 1999. *One Last Time*. New York: Berkley Books.

Gibson, Walter B. 1977. *The Original Houdini Scrapbook*. New York: Corwin/Sterling.

Gresham, William Lindsay. 1953. *Monster Midway*. New York: Rinehart.

Houdini, Harry. 1924. *A Magician Among the Spirits*. New York: Harper & Brothers.

Hyman, Ray. 1977. Cold reading: How to convince strangers that you know all about them. *Skeptical Inquirer* 1, no. 2 (Spring/Summer): 18–37.

Jaroff, Leon. 2001. Talking to the dead. *Time*, 5 March, 52.

Keene, M. Lamar. [1976] 1997. *The Psychic Mafia*. Amherst, N.Y.: Prometheus Books.

King, Larry. 2001. Are psychics for real? *Larry King Live*, aired 6 March.

McHargue, Georgess. 1972. *Facts, Frauds, and Phantasms: A Survey of the Spiritualist Movement*. Garden City, N.Y.: Doubleday.

Nickell, Joe. 1998. Investigating spirit communications. *Skeptical Briefs* (September): 5–6.

———. 1999. The Davenport Brothers: Religious practitioners, entertainers, or frauds? *Skeptical Inquirer* 23, no. 4 (July/August): 14–17.

———. 2000. Hustling Heaven. *Skeptical Briefs* 10, no. 3 (September): 1–3.

Nickell, Joe, with John F. Fischer. 1988. *Secrets of the Supernatural.* Buffalo, N.Y.: Prometheus Books, 47–60.

Nordlander, Charles [executive producer of *Crossing Over*]. 2001. Letter to the editor. *Time*, 26 March.

Schwartz, Gary E. R., et al. 2001. Accuracy and replicability of anomalous after-death communication across highly skilled mediums. *Journal of the Society for Psychical Research* (January): 1–25.

Wiseman, Richard, and Ciarán O'Keeffe. 2001. A critique of Schwartz et al.'s after-death communication studies. *Skeptical Inquirer* 25, no. 6 (November/December): 26–30.

22

Scandals and Follies of the "Holy Shroud"

The Shroud of Turin continues to be the subject of media presentations treating it as so mysterious as to imply a supernatural origin. One study (Binga 2001) found only 10 scientifically credible skeptical books on the topic, compared with more than 400 promoting the cloth as the authentic, or potentially authentic, winding sheet of Jesus—including, most recently, a revisionist tome entitled *The Resurrection of the Shroud* (Antonacci 2000). Since the cloth appeared in the middle of the fourteenth century, it has been at the center of scandal, exposés, and controversy—a dubious legacy for what is purported to be the holiest relic in Christendom.

Faked Shrouds

There have been numerous "true" shrouds of Jesus—along with vials of his mother's breast milk, hay from the manger in which he was laid after birth, and countless relics of his crucifixion—but the Turin cloth uniquely bears the apparent imprints of a crucified man. Unfortunately, the cloth is incompatible with New Testament accounts of Jesus' burial. John's gospel (19:38–42, 20:5–7) specifically states that the body was "wound" with "linen clothes" and a large quantity of burial spices (myrrh and aloes). Still another cloth (called "the napkin") covered his face and head. In contrast, the Shroud of Turin constitutes a *single*, *draped* cloth (laid under and then over the "body") without any trace of burial spices.

There were many earlier purported shrouds of Christ, which were typically about half the length of the Turin cloth. One was the subject of a reported seventh-century dispute, on the island of Iona, between Christians and Jews, both of whom claimed it. As adjudicator, an Arab ruler placed the alleged relic in a fire from which it levitated, unscathed, and fell at the feet of the Christians—or so says a pious tale. In medieval Europe alone, there were "at least forty-three 'True Shrouds'" (Humber 1978, 78).

Scandal at Lirey

The cloth now known as the Shroud of Turin first appeared about 1355 at a little church in Lirey, in north central France. Its owner, a soldier of fortune named Geoffroy de Charney, claimed it as the authentic shroud of Christ, although he never explained how he had acquired such a fabulous possession. According to a later bishop's report, written by Pierre D'Arcis to the Avignon pope, Clement VII, in 1389, the shroud was being used as part of a faith-healing scam:

> The case, Holy Father, stands thus. Some time since in this diocese of Troyes the dean of a certain collegiate church, to wit, that of Lirey, falsely and deceitfully, being consumed with the passion of avarice, and not from any motive of devotion but only of gain, procured for his church a certain cloth cunningly painted, upon which by a clever sleight of

> hand was depicted the twofold image of one man, that is to say, the back and the front, he falsely declaring and pretending that this was the actual shroud in which our Savior Jesus Christ was enfolded in the tomb, and upon which the whole likeness of the Savior had remained thus impressed together with the wounds which He bore. . . . And further to attract the multitude so that money might cunningly be wrung from them, pretended miracles were worked, certain men being hired to represent themselves as healed at the moment of the exhibition of the shroud.

D'Arcis continued, speaking of a predecessor who conducted the investigation and uncovered the forger: "Eventually, after diligent inquiry and examination, he discovered the fraud and how the said cloth had been cunningly painted, *the truth being attested by the artist who had painted it*, to wit, that it was a work of human skill and not miraculously wrought or bestowed" (emphasis added).

After this initial revelation, action had been taken and the cloth hidden away, but years later it had resurfaced. D'Arcis (1389) spoke of "the grievous nature of the scandal, the contempt brought upon the Church and ecclesiastical jurisdiction, and the danger to souls." As a consequence, Clement ordered that, while the cloth could continue to be exhibited (it had been displayed on a high platform flanked by torches), during the exhibition it must be loudly announced that "it is not the True Shroud of Our Lord, but a painting or picture made in the semblance or representation of the Shroud" (Humber 1978, 100). Thus the scandal at Lirey ended—for a time.

Further Misrepresentation

During the Hundred Years' War, Margaret de Charney, granddaughter of the Shroud's original owner, gained custody of the cloth, allegedly for safekeeping. Despite many subsequent entreaties, she refused to return it, instead even taking it on tour in the areas of present-day France, Belgium, and Switzerland. When there were additional challenges to the Shroud's authenticity, Margaret could only produce documents officially labeling it a "representation."

In 1453, at Geneva, Margaret sold the cloth to Duke Louis I of Savoy. Some Shroud proponents like to say Margaret "gave" the cloth to Duke Louis, but it is only fair to point out that in return he "gave" Margaret the sum of two castles. In 1457, after years of broken promises to return the cloth to the canons of Lirey and later to compensate them for its loss, Margaret was excommunicated. She died in 1460.

The Savoys (who later comprised the Italian monarchy and owned the shroud until it was bequeathed to the Vatican in 1983) represented the shroud as genuine. They treated it as a "holy charm" that had magical powers and enshrined it in an expanded church at their castle at Chambéry. There, in 1532, a fire blazed through the chapel, and before the cloth was rescued a blob of molten silver from the reliquary burned through its 48 folds. The alleged talisman was thus revealed as unable even to protect itself. Eventually, in a shrewd political move—by a later duke who wished a more suitable capital—the cloth was transferred to Turin (in present-day Italy).

In 1898 the shroud was photographed for the first time, and the glass-plate negatives showed a more lifelike, quasi-positive image (Figure 22-1). Thus began the modern era of the shroud, with proponents asking how a mere medieval forger could have produced a perfect "photographic" negative before the development of photography. In fact, the analogy with photographic images is misleading: the "positive" image shows a figure with white hair and beard, the opposite of what would be expected for a Palestinian Jew in his thirties.

Nevertheless, some shroud advocates suggested the image was produced by simple contact with bloody sweat or burial ointments. That theory, however, is disproved by a lack of wraparound distortions. Also, not all of the areas imaged would have been touched by a simple

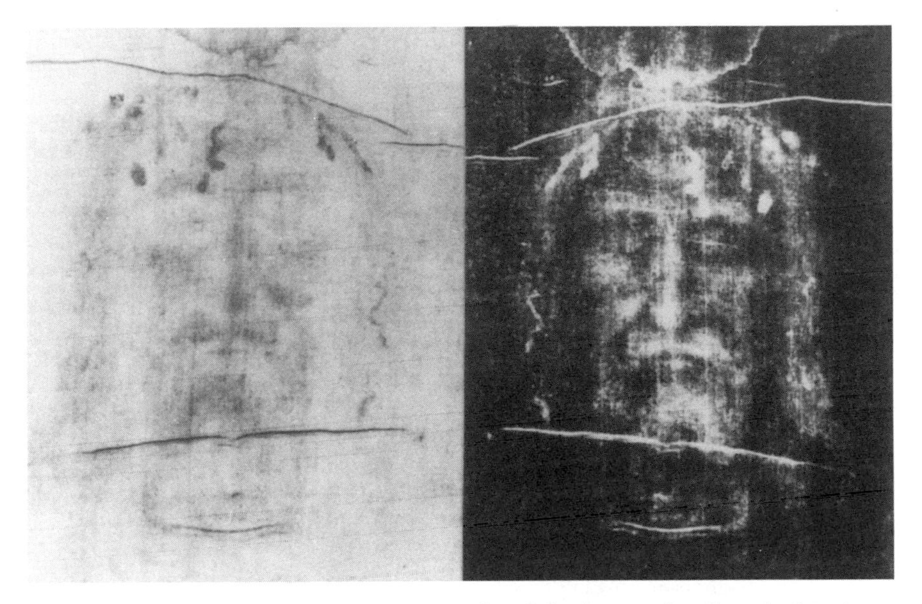

FIGURE 22-1. **Positive and negative photographs of the face on the Shroud of Turin. A medieval artist reported that the shroud was his handiwork, and scientific analyses confirm the presence of tempera paint.**

draped cloth, so some sort of *projection* was envisioned. One notion was "vaporography," body vapors supposedly interacting with spices on the cloth to yield a vapor "photo," but all subsequent experimentation produced only a blur (Nickell 1998, 81–84; except as otherwise noted, information is taken from this source). Others began to opine that the image had been "scorched" by a miraculous burst of radiant energy at the time of Jesus' resurrection. However, no known radiation would produce such superficial images, and actual scorches on the cloth from the fire of 1532 exhibit strong reddish fluorescence, in contrast to the shroud images which do not fluoresce at all.

Secret Commission

In 1969, the Archbishop of Turin appointed a secret commission to examine the shroud. That fact was leaked, and then denied, but (accord-

FIGURE 22-2. **Negative photograph of an image the author produced by making a rubbing from a bas-relief. Such a technique (using pigment or paint) automatically converts the usual lights and darks into a quasi-negative, shroudlike picture.**

ing to Wilcox 1977, 44) "[a]t last the Turin authorities were forced to admit what they previously denied." The man who exposed the secrecy accused the clerics of acting "like thieves in the night." More detailed studies—again clandestine—began in 1973.

The commission included internationally known forensic serologists who made heroic efforts to validate the "blood," but all of the microscopical, chemical, biological, and instrumental tests were negative. This was not surprising, as the stains were suspiciously still red and artistically "picturelike." Experts discovered reddish granules that would not even dissolve in reagents that dissolve blood, and one investigator found traces of what appeared to be paint. An art expert concluded that the image had been produced by an artistic printing technique (see Figure 22-2).

The commission's report was withheld until 1976 and then was largely suppressed, whereas a *rebuttal* report was made freely available. Thus began an approach that would be repeated over and over: distinguished experts were asked to examine the cloth, and then were attacked when they obtained other than the desired results.

Science versus "Shroud Science"

Further examinations were conducted in 1978 by the Shroud of Turin Research Project (STURP). STURP was a group of mostly religious believers whose leaders served on the Executive Council of the Holy Shroud guild, a Catholic organization that advocated the "cause" of the supposed relic. STURP members, like others calling themselves "sindonologists" (i.e., shroudologists), gave the impression that they started with the desired answer.

STURP pathologist Robert Bucklin—another Holy Shroud Guild executive councilman—stated that he was willing to stake his reputation on the shroud's authenticity. He and other pro-shroud pathologists argued for the anatomical correctness of the image—yet a footprint on the cloth is inconsistent with the position of the leg to which it is attached, the hair falls as for a standing rather than a recumbent figure, and the physique is so unnaturally elongated (similar to figures in Gothic art) that one pro-shroud pathologist concluded that Jesus must have suffered from Marfan's syndrome (Nickell 1989)!

STURP lacked experts in art and forensic chemistry—with one exception: famed microanalyst Walter C. McCrone. Examining 32 samples tape-lifted from the shroud, McCrone identified the "blood" as tempera paint containing red ocher and vermilion along with traces of rose madder—pigments used by medieval artists to depict blood. He also discovered that on the image—but not the background—were significant amounts of the red ocher pigment. He first thought that this had been applied as a dry powder, but later concluded that it was a component of dilute paint applied in the medieval *grisaille* (monochromatic) technique (McCrone 1996; cf. Nickell 1998). For his efforts McCrone was held to a secrecy agreement, while statements made to the press indicated that there was no evidence of artistry. McCrone was then, he says, "drummed out" of STURP.

STURP representatives paid a surprise visit to McCrone's lab to confiscate his samples, then gave them to two late additions to STURP, John Heller and Alan Adler, neither of whom was a forensic serologist or a pigment expert. The pair soon proclaimed they had "identified the presence of blood." However, at the 1983 conference of the prestigious International Association for Identification, forensic analyst John F. Fischer explained how results similar to theirs could have been obtained from tempera paint.

A more recent claim concerns reported evidence of human DNA in a shroud "blood" sample, although the Archbishop of Turin and the Vatican refused to authenticate the samples or accept any research carried out on them. University of Texas researcher Leoncio Garza-Valdez, in his *The DNA of God?* (1999, 41), claims that it was possible "to clone the sample and amplify it," thereby proving that it was "ancient" blood "from a human being or high primate"; Ian Wilson's *The Blood and the Shroud* (1998, 91) flatly asserted that it was "human blood."

Actually, the scientist at the DNA lab, Victor Tryon, told *Time* magazine that he could not say how old the DNA was or confirm that it came from blood. As he explained, "Everyone who has ever touched the shroud or cried over the shroud has left a potential DNA signal there." Tryon resigned from the new shroud project due to what he disparaged as "zealotry in science" (Van Biema 1998, 61).

Pollen Fraud?

McCrone would later refute another bit of pro-shroud propaganda: the claim of a Swiss criminologist, Max Frei-Sulzer, that he had found certain pollen grains on the cloth that "could only have originated from plants that grew exclusively in Palestine at the time of Christ." Earlier Frei had also claimed to have discovered pollens on the cloth that were characteristic of Istanbul (formerly Constantinople) and the area of ancient Edessa—seeming to confirm a "theory" of the shroud's missing early history. Wilson (1979) conjectured that the shroud was the fourth-century Image of Edessa, a legendary "miraculous" imprint of Jesus' face made as a gift to King Abgar. Wilson's notion was that the shroud had been folded so that only the face showed and that it had thus been disguised for centuries. Actually, had the cloth been kept in a frame for such a long period, there would have been an age-yellowed, rectangular area around the face. Nevertheless, Frei's alleged pollen evidence gave new support to Wilson's ideas.

I say *alleged* evidence because Frei had severe credibility problems. Before his death in 1983, his reputation suffered greatly when, after representing himself as a handwriting expert, he pronounced the infamous "Hitler diaries" genuine. They were soon exposed as forgeries.

In the meantime, an even more serious question had arisen about Frei's pollen evidence. Although he reported finding numerous types of pollen from Palestine and other areas, STURP's tape-lifted samples, taken at the same time, showed little pollen. Micropaleontologist Steven D. Schafersman was probably the first to publicly suggest that Frei might be guilty of deception. He explained how unlikely it was, given the evidence of the shroud's exclusively European history, that 33 different Middle Eastern pollens could have reached the cloth—particularly only pollen from Palestine, Istanbul, and the Anatolian steppe. With such selectivity, Schafersman stated, "these would be miraculous winds indeed." In an article in *Skeptical Inquirer*, Schafersman (1982) called for an investigation of Frei's work.

When Frei's tape samples became available after his death, McCrone was asked to authenticate them. This he was readily able to do, he told me, "since it was easy to find red ocher on linen fibers much the same as I had seen them on my samples." But there were few pollen grains other than on a single tape, which bore "dozens" in one small area. This indicated that the tape had subsequently been "contaminated," probably deliberately, McCrone concluded, by having been pulled back and the pollen surreptitiously introduced.

McCrone added (1993):

> One further point with respect to Max which I haven't mentioned anywhere, anytime to anybody is based on a statement made by his counterpart in Basel as head of the Police Crime Laboratory there that Max had been several times found guilty and was censured by the Police hierarchy in Switzerland for, shall we say, overenthusiastic interpretation of his evidence. His Basel counterpart had been on the investigating committee and expressed surprise in a letter to me that Max was able to continue in his position as Head of the Police Crime Lab in Zurich.

C-14 Falsehoods

The pollen "evidence" became especially important to believers following the devastating results of radiocarbon dating tests in 1988. Three laboratories (at Oxford, Zurich, and the University of Arizona) used accelerator mass spectrometry (AMS) to date samples of the linen. The results were in close agreement and were given added credibility by the use of control samples of known dates. The resulting age span was circa C.E. 1260–1390—consistent with the time of the reported forger's confession.

Shroud enthusiasts were devastated, but they soon rallied, beginning a campaign to discredit the radiocarbon findings. Someone put out a false story that the AMS tests were done on one of the patches from the 1532 fire, thus supposedly yielding a late date. A Russian scientist, Dmitrii Kuznetsov, claimed to have established experimentally that heat from a fire (like that of 1532) could alter the radiocarbon date. Others could not replicate his alleged results, however, and it turned out that his physics calculations had been plagiarized—complete with an error (Wilson 1998, 219–223). (Kuznetsov was also exposed in *Skeptical Inquirer* for bogus research in a study criticizing evolution [Larhammar 1995].)

A more persistent challenge to the radiocarbon testing was hurled by Garza-Valdez (1999). He claimed to have obtained a swatch of the "miraculous cloth" that bore a microbial coating, contamination that could have altered the radiocarbon date. That notion was effectively debunked by physicist Thomas J. Pickett (1996). He performed a simple calculation showing that, for the shroud's date to have been altered by 13 centuries (i.e., from Jesus' first-century death to the radiocarbon date of 1325±65 years), there would have had to be twice as much contamination, by weight, as the cloth itself!

Shroud of Rorschach

Following the suspicious pollen evidence were claims that plant images had been identified on the cloth. These were allegedly discerned from "smudgy"-appearing areas in shroud photos that were subsequently enhanced. The work was done by a retired geriatric psychiatrist, Alan Whanger, and his wife Mary, former missionaries who took up image analysis as a hobby. They were later assisted by an Israeli botanist who looked at their photos of "flower" images (many of them "wilted" and otherwise distorted) and exclaimed, "Those are the flowers of Jerusalem!" Apparently no one has thought to see if some might match the flowers of France or Italy or even to try to prove that the images are indeed floral (given the relative scarcity of pollen grains on the cloth).

The visualized "flower and plant images" join other shapes perceived, Rorschach-style, in the shroud's mottled image and off-image areas. These include "Roman coins" over the eyes, head and arm "phylacteries" (small Jewish prayer boxes), an "amulet," and such crucifixion-associated items (cf. John, ch. 19) as "a large nail," a "hammer," "sponge on a reed," "Roman thrusting spear," "pliers," "two scourges," "two brush brooms," "two small nails," "large spoon or trowel in a box," "a loose coil of rope," a "cloak" with "belt," a "tunic," a pair of "sandals," and other far-fetched imaginings, including "Roman dice"—all discovered by the Whangers (1998) and their botanist friend.

They and others have also reported finding ancient Latin and Greek words, such as "Jesus" and "Nazareth." Even Ian Wilson (1998, 242) felt compelled to state: "While there can be absolutely no doubting the sincerity of those who make these claims, the great danger of such arguments is that researchers may 'see' merely what their minds trick them into thinking is there."

Conclusion

We see that "shroud science"—like "creation science" and other pseudosciences in the service of dogma—begins with the desired answer and works backward to the evidence. Although they are bereft of any viable hypothesis for the image formation, sindonologists are quick to dismiss the profound, corroborative evidence for artistry. Instead, they suggest that the "mystery" of the shroud implies a miracle, but of course that is merely an example of the logical fallacy called arguing from ignorance.

Worse, some have engaged in pseudoscience and even, apparently, outright scientific fraud, while others have shamefully mistreated the honest scientists who reported unpopular findings. We should again recall the words of Canon Ulysse Chevalier, the Catholic scholar who brought to light the documentary evidence of the shroud's medieval origin. As he lamented, "The history of the shroud constitutes a protracted violation of the two virtues so often commended by our holy books: justice and truth."

REFERENCES

Antonacci, Mark. 2000. *The Resurrection of the Shroud: New Scientific, Medical and Archeological Evidence.* New York: M. Evans.

Binga, Timothy. 2001. Report in progress from the Director of the Center for Inquiry Libraries, 19 June.

D'Arcis, Pierre. [1389] 1979. Memorandum to the Avignon Pope, Clement VII. Translated from Latin by Rev. Herbert Thurston; Reprinted in *The Shroud of Turin*, by Ian Wilson, 266–72. Rev. ed. Garden City, N.Y.: Image Books.

Garza-Valdez, Leoncio A. 1993. *Biogenic Varnish and the Shroud of Turin*. Cited in Garza-Valdez 1999, 37.

———. 1999. *The DNA of God?* New York: Doubleday.

Humber, Thomas. 1978. *The Sacred Shroud*. New York: Pocket Books.

Larhammar, Dan. 1995. Severe flaws in scientific study criticizing evolution. *Skeptical Inquirer* 19, no. 2 (March/April): 30–31.

McCrone, Walter C. 1993. Letters to Joe Nickell, 11 June and 30 June.

———. 1996. *Judgement Day for the Turin "Shroud."* Chicago: Microscope Publications.

Nickell, Joe. 1989. Unshrouding a mystery: Science, pseudoscience, and the Cloth of Turin. *Skeptical Inquirer* 13, no. 3 (Spring): 296–99.

———. 1998. *Inquest on the Shroud of Turin: Latest Scientific Findings*. Amherst, N.Y.: Prometheus Books.

Pickett, Thomas J. 1996. Can contamination save the Shroud of Turin? *Skeptical Briefs* (June): 3.

Schafersman, Steven D. 1982. Science, the public and the Shroud of Turin. *Skeptical Inquirer* 6, no. 3 (Spring): 37–56.

Van Biema, David. 1998. Science and the shroud. *Time*, 20 April, 53–61.

Whanger, Mary, and Alan Whanger. 1998. *The Shroud of Turin: An Adventure of Discovery*. Franklin, Tenn.: Providence House.

Wilcox, Robert K. 1977. *Shroud*. New York: Macmillan.

Wilson, Ian. 1979. *The Shroud of Turin*. Rev. ed.. Garden City, N.Y.: Image Books.

———. 1998. *The Blood and the Shroud*. New York: Free Press.

23

"Pyramid Power" in Russia

Still the largest country in the world, Russia retains more than 76 percent of the area of the former USSR, which collapsed in 1991. The collapse, along with the suspension of activities of the Communist Party, increased *glasnost* ("openness") in the new federal republic. With personal freedom, however, has come a rise in pseudoscientific and magical expression.

I became increasingly aware of this through the visits of three Russian notables to the Center for Inquiry–International: Valerií Kuvakin (professor of philosophy at Moscow State University), Edward Kruglyakov (a distinguished physicist at Novosibirsk, Siberia), and Yurií Chornyi (Scientific Secretary, Institute for Scientific Information, Russian Academy of Sciences). Subsequently, I was one of several CSICOP speakers at an international congress, "Science, Antiscience, and the

Paranormal," held in Moscow (October 3–5, 2001) and co-sponsored by the Russian Academy of Sciences. There I learned more about the public's newfound glasnost toward all things mysterious. I stayed on for several more days in order to investigate some of these. (See also chapters 30, 37, and 39.)

Psychic Discoveries

If it is true that the mystically oriented New Age movement began about 1971 (despite having its roots in earlier periods) (Melton 1996), perhaps its most immediate harbinger was a book by Sheila Ostrander and Lynn Schroeder, *Psychic Discoveries Behind the Iron Curtain* (1970). Its claims, which dominated interest in the paranormal for years, included many that remain familiar (and continue to provoke skeptical response).

There was Russian (via Chinese) acupuncture, a form of medical therapy that supposedly influenced the flow of *qi* (pronounced *chee*) or "life force" (although skeptics suspect that the beneficial, pain-relief effects are largely due to the body's production of narcotic-like chemicals called *endorphins* [Barrett 1996, 18]). There was also Kirlian photography, a noncamera technique in which a high-voltage, high-frequency electrical discharge is applied to a grounded object (such as a leaf, finger, etc.) to yield an "aura" that can be recorded on a photographic plate, film, or paper. (Actually, the supposed aura is only "a visual or photographic image of a corona[l] discharge in a gas, in most cases the ambient air," and the result is influenced by moisture, finger pressure, and other physical factors [Watkins and Bickel 1986].)

Psychic Discoveries Behind the Iron Curtain also touted dowsing (which has failed repeated scientific tests elsewhere and is attributed to unconscious muscular activity [Vogt and Hyman 1979]); demonstrations of psychokinesis and dermo-optical perception (accomplished by Russian ladies using simple conjuring tricks [Randi 1995, 40]); and other fanciful, and now discredited, "psychic" topics, including "pyramid power."

Pyramid Power

In the wake of the new glasnost, pyramids sprang up across the Russian landscape. These are a modern expression of a craze, fostered by *Psychic Discoveries Behind the Iron Curtain*, involving "the secrets of the pyramids." Citing the Great Pyramid of Cheops in Egypt—"one of the seven wonders of the world and one of the strangest works of architecture in existence"—the authors claimed that small cardboard models of the Cheops pyramid could preserve food (especially "mummify meat"), relieve headaches, sharpen razor blades, and possibly perform other wonders (Ostrander and Schroeder 1970, 366–76).

The specific claims came from a Czechoslovakian radio engineer, Karel Drbal, who obtained Czech patent 91304 for the Cheops Pyramid Razor-Blade Sharpener. It supposedly generated some unknown and mysterious "energy." Unfortunately, what Ostrander and Schroeder seem to mean by that term is *mystical power*, but they call it "energy" as if to give it a semblance of scientific legitimacy.

Alas, subsequent tests of the claims failed to substantiate them. Pyramids preserved organic matter no better than containers of other shapes did; nor did placing razor blades in pyramids restore the blades' sharpness, despite the subjective judgments of people fooled by their own expectations (Hines 1988, 306–7). Ostrander and Schroeder (1970, 372–73) tried pyramid power on their own razor blades and thus were able to "attest" that the "blades sharpened up again and remained sharp if kept in the pyramid between uses," but they made no mention of using controls (i.e., other shapes of containers for comparison of results), scientific methods of determining sharpness, or double-blind procedures.

Nevertheless, boosted by claims that pyramid power had been unleashed "behind the iron curtain," the pyramid craze flourished in the United States. One company marketed a kit with eight wooden sticks that formed a pyramidal frame when glued in place. Even without being covered (say, with paper or foil), this little structure could allegedly retard food spoilage, remove the bitterness from coffee, impart a mellower taste to wine, sharpen razor blades, perk up houseplants, and perform other wonders—according to the kit's guide book, that is, which claims, "The Pyramid is a geometric focusing lens of cosmic energy" (Kerrell and Goggin 1974; see also Toth and Nielsen 1985, 139–50).

A similar wire-frame "magic pyramid" was made to be worn on the head (where it looked rather like a dunce cap) in order to concentrate the wearer's alleged psychic or healing powers (Randi 1982, 206–7). Larger frames were available for one to sit inside in the lotus position as a means of improving meditation, or to place over the bed to gain enhanced vitality (Kerrell and Goggin 1974, 6–7). There were even pyramid-shaped doghouses that supposedly rid their occupants of fleas (Hines 1988).

The pyramid craze lasted through the 1970s (Randi 1995, 194) and then declined, although it has never entirely gone away. New Age merchants still offer small gemstone pyramids that focus the "energy" of the particular stone (e.g., tigereye for enhancement of "psychic abilities"), as well as plastic "Wishing Pyramids" (into which is placed a paper with one's wish written thereon), and other items.

In Russia, pyramid power is on the rise—almost literally: I was able to visit one that towered 44 meters (about 144.4 feet or some 12 stories). Built in 1999 by Alexander Golod, it is the tallest of about 20 such pyramids intended for alleged scientific and medical purposes. I was taken to the site—about 38 kilometers northwest of Moscow—by Valerií Kuvakin (who also translated various interviews) and his wife, Uliya Senchihina (see Figure 23-1). Valerií heads the Center for Inquiry–Moscow.

Although it resembles a stone structure from the outside, from the interior the pyramid is seen to be constructed of translucent Plexiglas panels over a wood framework. It was closed when we arrived (late one afternoon), but a custodian consented to let us in and show us around.

FIGURE 23-1. **Valerií Kuvakin and his wife, Uliya Senchihina, approach a pyramid northwest of Moscow. It is one of several that dot the Russian countryside and supposedly utilize "pyramid power" for various beneficial purposes.**

The pyramid was largely empty, although off to one side were cases of bottled water that were supposedly being energized for curative purposes.

A rope-cordoned central area, where the pyramid power is supposedly most concentrated, contained some crystal spheres that were also being "energized." Despite being warned that the energy there was so intense that someone with a large "biofield" (or "aura")—such as I supposedly have—could lose consciousness, I ducked under the rope to dare the awesome power. I stood there for a time (while Valerií photographed me, barely containing his amusement, and while we continued our conversation with the custodian), but I felt no effect whatsoever.

Adjacent to the pyramid, in a small outbuilding, was a small gift shop. A sign advertised (in translation) "Consumer Goods Energized by Pyramid." I decided to forgo the "energized" gemstones: A woman at a store where we stopped en route for directions said she had bought some of the stones, which were supposed to cure headaches, "and they didn't work," she stated, disparaging the pyramid claims. I did buy a small "energized" lead-crystal (glass) pyramid and a booklet titled (in translation), *Pyramids of the Third Millennium*. It featured the various new pyramids, including the one we were then visiting.

According to this source, although the claims for pyramid power might seem like mysticism and shamanism to some, those with "intuition" would see in them the basis of a new physics, a new biology, and so on. It boasted that the new pyramids have reduced the incidence of cancer, AIDS, and other diseases in the areas surrounding them; have begun to cleanse their local environments; and promise to extend nearby residents' longevity to more than 100 years. The booklet even promised that pyramids could help reduce religious and other conflicts (although we might note that they have not seemed effective in that regard, historically, in Egypt).

Valeriĭ noted that the claims were published without any supporting scientific data. Physicist Edward Kruglyakov (2001), who had previously visited the site and suggested it to me, found the claims utterly outlandish, unsubstantiated, and scientifically without merit.

REFERENCES

Barrett, Stephen. 1996. "Alternative" health practices and quackery. In *The Encyclopedia of the Paranormal*, by Gordon Stein, ed. Amherst, N.Y. Prometheus Books, 18.

Hines, Terence. 1988. *Pseudoscience and the Paranormal*. Buffalo, N.Y.: Prometheus Books.

Kerrell, Bill, and Kathy Goggin. 1974. *Basic Pyramid Experimental Guide Book*. Santa Monica, Cal.: Pyramid Power-V, Inc.

Kruglyakov, Edward. 2001. Personal communications, 6 April and 3–5 October.

Melton, J. Gordon, ed. 1996. *Encyclopedia of Occultism & Parapsychology*. Detroit, Mich.: Gale Research, s.v. "New Age" (vol. II, pp. 922–24).

Ostrander, Sheila, and Lynn Schroeder. 1970. *Psychic Discoveries Behind the Iron Curtain*. New York: Bantam Books.

Randi, James. 1982. *Flim-Flam! Psychics, ESP, Unicorns and Other Delusions*. Buffalo, N.Y.: Prometheus Books.

———. 1995. *An Encyclopedia of Claims, Frauds, and Hoaxes of the Occult and Supernatural*. New York: St. Martin's Griffin.

Toth, Max, and Greg Nielsen. 1985. *Pyramid Power*. Rochester, Vt.: Destiny Books.

Vogt, Evon Z., and Ray Hyman. 1979. *Water Witching U.S.A.* 2d ed. Chicago: University of Chicago Press.

Watkins, Arleen J., and William S. Bickel. 1986. A study of the Kirlian effect. *Skeptical Inquirer* 10, no. 3 (Spring): 244–57.

24

Diagnosing the "Medical Intuitives"

Among the dangerous new, pseudoscientific fads is that of "energy medicine," practiced by self-styled "medical intuitives." Actually, the practice is not new but only newly resurging, like most other aspects of the so-called New Age movement. It involves using some form of reputed psychic ability to divine people's illnesses and, typically, recommend treatment for them.

The best-known exponent of this field was Edgar Cayce, who flourished in the first half of the twentieth century and offered medical diagnoses while hypnotized. He had many predecessors, though, notably the "Poughkeepsie Seer," Andrew Jackson Davis (1826–1910), a forerunner of modern spiritualism who similarly diagnosed while in a supposed mesmeric trance (Doyle 1926, 42–59). Many other Cayce predecessors practiced in the heyday of spiritualism, as shown by advertisements in spiritualist journals. For example, one such publication included ads for a "Medical Clairvoyant," a "Healing Mesmerist," a "Clairvoyant, and Healing Medium" (who also diagnosed by correspondence), and the like. An "Electro-Magnetic Healer" worked via his "Spirit-Physicians" to become "a Powerful Healing Medium," and a married couple of "Medical Clairvoyantes" offered "Medical Diagnosis by Lock of Hair" (*Medium* 1875).

Among Cayce's more immediate predecessors was a now almost forgotten lady from western New York, Mrs. Antoinette Matteson. Here is her story, followed by a look at Cayce and then the modern medical intuitives. As we shall see, despite some differences, they represent a certain type of paranormal claimant who is prone to fantasizing.

"Clairvoyant Doctress"

Several years ago, I purchased from an antique dealer an old blown-in-a-mold medicine bottle embossed, "MRS. J. H. R. MATTESON. / CLAIRVOYANT REMEDIES. / BUFFALO, N.Y." Later, from the same dealer, I acquired another, embossed "CLAIRVOYANT" and "PSYCHIC REMEDIES." I believe the latter was an earlier one of Mrs. Matteson's, but who was she? Well, as I discovered, she was a character.

Born Antoinette Wealthy in Baden, Germany, in 1847, she came with her parents to the United States at the age of five. In 1857, the family moved from Water Valley, New York, to Buffalo. "At an early age," she found herself, she said, "subordinate to the control of certain occult influences" or "Spirit Vision" which "beholds and proclaims what the material eye does not and cannot see""(Matteson 1894, 9).

In 1864, Antoinette married Judah H. R. Matteson, a musician. Although blind, he "had remarkable ability in getting about the country alone" (Whitcomb 1923). He died in 1884 of "congestion of the brain," possibly resulting from "the excessive use of liquor." According to his obituary, his wife, by then "a well-known clairvoyant," awoke that Sunday morning, and assumed that her husband was still asleep. "While the remainder of the family were at breakfast Mrs. Matteson, as she claims, went into a clairvoyant state, and while in that condition announced to her children that her husband was dead"—as indeed he proved to be ("The singular death" 1884). (Note the newspaper's skepticism, evident in the phrase "as she claims.")

After her husband's death, Mrs. Matteson became the sole support of her family, and listed herself in the Buffalo City Directory as a "clairvoyant doctress." According to a local newspaper's retrospection:

> She had hosts of patrons and her practice was to consult with a patron, learn their ailments and symptoms and then pretend to go into a trance state, during which excitement she uttered the number or numbers for the patient under observation. Each number appertained to certain alleged remedies she manufactured and sold to her patrons. After "emerging" from her alleged trance, she asked for the numbers quoted and dispensed her remedies accordingly ["Mrs. Matteson" 1929].

Mrs. Matteson's remedies were herbal and she eventually self-published *The Occult Family Physician and Botanic Guide to Health* (1894) (see Figure 24-1). In it she criticized the "old school of Medicine" for rejecting that which is different, and advanced her "remedies from the vegetable world" together with other natural methods, including "the grander forces of the Spirit, Magnetic, Clairvoyance, Psychoma or Hypnotism, Electricity, Water cure, and also the power of sun-light, etc., which," she stated, "are beyond question, in advance of the old stereotyped process."

As well, despite "skeptical science," she touted spiritualism and insisted, "Short of spiritual manifestations by decarnated spirits, clairvoyance is one of the best scientific proofs of immortality we have." She

FIGURE 24-1. **Antoinette (Mrs. J. H. R.) Matteson, "clairvoyant doctress," from the frontispiece of her 1894 book, *The Occult Family Physician and Botanic Guide to Health.***

claimed, "During the twenty years of my mediumistic experience, many hundreds, in fact I may say thousands of remarkable cures have been made through the aid of my spirit guides."

In fact, Antoinette Matteson exhibited many of the traits associated with a fantasy-prone personality. Such people are sane and normal but have a propensity to fantasize. Typically, they are easily hypnotized, believe they have special psychic and healing powers, think that they receive special messages from higher entities, and so forth. They often turn up as spiritualist mediums, religious visionaries, UFO abductees, or other such types (Baker 1990, 245–50; Baker and Nickell 1992, 221–24).

During Mrs. Matteson's career, the Erie County Medical Board tried to end her "practice," and she was repeatedly arrested for "practicing medicine without a license." However, it is said no grand jury would ever indict her (Whitcomb 1923). Interestingly, one of her daughters managed to graduate from the Buffalo Medical College, despite hostility from the profession on account of her mother's identity ("Mrs. Matteson" 1929).

In 1910 Mrs. Matteson retired, turning over her business to another daughter, Mrs. Nellie Whitcomb, who was also a spiritualist and had been her mother's "assistant for thirty years." Antoinette "Passed to Higher Life" on 11 October 1913, and Nellie continued to act as "wholesale and Retail Agent for the Old Original Mrs. J. H. R. Matteson's Psychic Clairvoyant Remedies" for a number of years (Whitcomb 1923).

"Sleeping Prophet"

In February 2000, while in Virginia to lecture at the Old Dominion University in Norfolk, I had a chance to visit the Association for Research and Enlightenment (A.R.E.) in nearby Virginia Beach. Founded in 1927, it promotes the life work of Edgar Cayce, the so-called "Sleeping Prophet," who uttered prognostications and gave medical readings while supposedly hypnotized. The A.R.E. library includes bound volumes of transcripts of 14,306 Cayce readings.

Born in 1877 on a farm near Hopkinsville, Kentucky, Cayce was a "dreamy" child, especially prone to fantasies. For example, he had imaginary playmates. Although his schooling did not go past the ninth grade, he worked in a bookstore and read extensively—especially about mystical subjects, including what today would be called "alternative" medical practices. Cayce became a photographer and operated a studio in Hopkinsville, where he met a man named Al Layne. Layne had been thwarted in his desire to become a doctor and instead took cheap mail-order courses in hypnotism and osteopathy, a forerunner of chiropractic that emphasizes "manipulation" (special application of the healer's hands to joints, muscles, and so on).

Cayce soon discovered that he was a good hypnotic subject—he could even put himself into an alleged trance. In that condition, he gave people readings concerning their health, as he was apparently able to diagnose illnesses and prescribe effective remedies for them. He became known locally as a sort of eccentric folk doctor. Then, on 9 October 1910, the *New York Times* Sunday magazine ran an article that boosted his reputation. Headlined "Illiterate Man Becomes a Doctor When Hypnotized—Strange Power Shown by Edgar Cayce Puzzles Physicians," the article began: "The medical fraternity of the country is taking a lively interest in the strange power said to be possessed by Edgar Cayce of Hopkinsville, Ky., to diagnose difficult diseases while in a semiconscious state, though he has not the slightest knowledge of medicine when not in this condition."

Note the exaggerations in the article, beginning with the first word of the headline. Cayce was far from "illiterate," and in fact his various diagnoses suggest that he was heavily influenced by osteopathy and other quaint theories of healing, obtained from books and from his association with people such as Layne. Most of his early readings were given with the aid of Layne, the osteopath, who asked questions while Cayce was supposedly entranced. According to Cayce critic Martin Gardner (1957, 216–19): "There is abundant evidence that Cayce's early association with osteopaths and homeopaths had a major influence on the character of his readings. Over and over again he would find spinal lesions of one sort or another as the cause of an ailment and prescribe spinal manipulations for its cure."

In addition to osteopathic manipulations, Cayce prescribed electrical treatments, special diets, and various tonics and other remedies. For a baby with convulsions he prescribed "peach-tree poultice"; for a leg sore, something called "oil of smoke"; and for a priest with an epilepsy-like condition, the use of "castor oil packs." He offered several remedies for baldness, including giving the scalp a rubbing with crude oil and "Listerine twice a week." He thought almonds could cure can-

FIGURE 24-2. **The author studies Edgar Cayce readings at the Association for Research and Enlightenment library.**

cer, and for an ailment called dropsy he prescribed "bedbug juice" (Gardner 1957, Nickell 1991).

As I looked over some of the thousands of transcripts of "patient" case histories on file at the A.R.E. library (Figure 24-2), I was horror-struck by some of the prescriptions. For example, for poliomyelitis Cayce recommended a mixture of "½ gal. pure gasoline / + 1 oz. oil of cedar / 1 drm. camphor gum / [and] 1 oz. oil of sassafras." This was to be massaged, he directed, into "lumbar centers and the brachial centers, and over the cerebro-spinal centers themselves" (Cayce n.d.).

Such bizarre treatments do not inspire confidence, but the question remains: How could Cayce apparently diagnose people's illnesses from hundreds of miles away? James Randi (1982, 189) explains that "Cayce was fond of expressions like 'I feel that . . . ' and 'perhaps'—qualifying words used to avoid positive declarations." He adds: "Many of the letters he received—in fact, most—contained specific details about the illnesses for which readings were required, and there was nothing to stop Cayce from knowing the contents of the letters and presenting that information as if it were divine revelation." At times Cayce was hilariously inaccurate, providing diagnoses of subjects *who had died* since their letters had been sent! Unaware of the fact, Cayce simply rambled on in his usual fashion. For one, he prescribed an amazing mixture made from Indian turnip, wild ginseng, and other ingredients. As Randi says of Cayce's failure to know that these patients were deceased, "Surely, death is a very serious symptom, and should be detectable" (Randi 1982, 191–92).

As such incidents indicate, although he was touted as a seer, Cayce's prognosticatory ability left something to be desired. He failed to foresee his own arrest, in New York, for *fortunetelling!* When Pearl Harbor was attacked, Cayce was "as stunned as anybody else" (Stearn 1968, 16–17). Sometimes his forecasts were ridiculous: For instance, he predicted that Atlantis, the mythical sunken continent, would rise again in the 1960s. Of course, it did not (Nickell 1991).

Edgar Cayce died in 1945, apparently without foreseeing his own end. Those who still insist that he had some special powers come from the ranks of Cayce's fellow mystics. Even the A.R.E., in a book of extracts from the readings, cautions, "Application of medical information found in the Cayce readings should be undertaken only with the advice of a physician" (Turner and St. Clair 1976).

Today's Crop

The field of so-called medical intuition has grown with the New Age movement, and gained considerable attention in the 1990s with the publication of several books on the subject, including the best-selling *Why People Don't Heal and How They Can*. Written by Caroline Myss (1997), a Ph.D. "who has a background in theology" (Koontz 2000, 102), it claims that she and others can divine illness by some sort of psychic means—in Myss's case, by reading "the energy field that permeates and surrounds the body" (often called the *aura*) (Myss 1997, xi). Although Myss herself no longer gives readings, she is involved with a "holistic" physician, C. Norman Shealy, in a program to "certify" medical intuitives (Koontz 2000, 66, 102).

Another physician who touts—and also practices—medical intuition is Judith Orloff, M.D., a psychiatrist. She wrote *Second Sight* (1996), a book about her alleged psychic abilities and an unintentional demonstration that—typical of those in the field—she has many of the traits associated with fantasy-prone personalities. That first book was followed by *Dr. Judith Orloff's Guide to Intuitive Healing* (2000). She asks, "What could be more natural than a doctor with psychic insight who can heal not only with medicine but with energy?" (Orloff 1996, 352).

Alas, in contrast to objective, scientific medicine, which continues to make important breakthroughs in identifying and treating diseases, injuries, and other illnesses, "energy medicine" is based on mysticism and pseudoscience. Often incorporating such New Age fads as astrology, yoga, and reincarnation, it is part of what Myss terms "the alternative healing community." The "healers" offer varied procedures. For lupus, for example (an ulcerous skin disease that Myss admits has "essentially no cure"), the faddish healers may offer "treatments" ranging "from acupuncture to visualization to aromatherapy." For those who may be squeamish about acupuncture, there are such alternatives as acupressure (which foregoes the needles), reflexology (limited to pressure on certain zones of the feet), simple massage, the biblical "laying on of hands," or even "therapeutic touch" (a misnomer for merely passing the hands over the subject). In addition, there are "talk therapy," crystal healing, herbal remedies, homeopathy, meditation, and, of course, prayer. "Skilled physician" is listed as just one among many possibilities (Myss 1997, 102–103, 148).

Almost anything will do. Myss encourages individuals to believe: "There are no wrong choices. Every choice I believe in is an effective means of healing." Her caution to "get a second and even a third professional opinion" is weakened by her definition of *professional* to include virtually all types of "holistic" practitioners. "Any treatment," she states, "that can enhance your healing and bring hope and strength back to your body is worth considering" (Myss 1997, x, 144, 159). Nowhere does Myss cite any scientific double-blind experiments in support of such alternative treatments. Instead, she merely offers the old feel-good remedies of "spirituality," the power of positive thinking, and the placebo effect.

Although the intuitives do not dismiss "conventional" medicine, their advocacy of "alternative" and "complementary" treatments may lead desperate seekers to just such a dismissal, with potentially tragic consequences. That is why I told *New Age* magazine that medical intuition was not only pseudoscientific but, in a word, "dangerous" (Koontz 2000, 102).

Caroline Myss's philosophy is that "[o]ur lives are made up of a series of mysteries that we are meant to explore but that are meant to remain unsolved." Such mystery-mongering naturally leads to occult, mystical, and magical thinking. A more enlightened view would hold that mysteries should be neither fostered nor suppressed, but rather carefully investigated in hopes of solving them. Indeed, one can see the progress of civilization as a series of solved mysteries. This is the attitude that led to the development of polio vaccine and the eradication of smallpox. "Energy medicine" can boast of no comparable successes.

REFERENCES

Baker, Robert A. 1990. *They Call It Hypnosis*. Buffalo, N.Y.: Prometheus Books.

Baker, Robert A., and Joe Nickell. 1992. *Missing Pieces: How to Investigate Ghosts, UFOs, Psychics, & Other Mysteries*. Buffalo, N.Y.: Prometheus Books.

Cayce, Edgar. N.d. Readings (A.R.E. Library), vol. 10, no. 135–1, M. 24, p. 2.

Doyle, Arthur Conan. [1926] 1975. *The History of Spiritualism*. Reprinted New York: Arno Press.

Gardner, Martin. 1957. *Fads and Fallacies in the Name of Science*. New York: Dover.

Illiterate man becomes a doctor when hypnotized. 1910. *New York Times* (Sunday magazine), 9 October.

Koontz, Katy. 2000. The new health detectives. *New Age* (January/February): 64–66, 102–110.

Matteson, Antonette [sic]. 1894. *The Occult Family Physician and Botanic Guide to Health*. Buffalo, N.Y.: Privately printed [copy in author's collection].

The Medium and Daybreak. 1875. London, 13 August, 527.

Mrs. Matteson, Clairvoyant. 1929. *The Times* (Buffalo, N.Y.), 14 February. (Clipping in "Local Biographies" scrapbook, Special Collections, Buffalo and Erie County Public Library, vol. 22, p. 306.)

Myss, Caroline. 1997. *Why People Don't Heal and How They Can*. New York: Harmony Books.

Nickell, Joe. 1991. "The Sleeping Prophet": Edgar Cayce. In *Wonderworkers! How They Perform the Impossible*, by Joe Nickell, 76–83. Buffalo, N.Y.: Prometheus Books.

Orloff, Judith. 1996. *Second Sight*. New York: Warner Books.

———. 2000. *Dr. Judith Orloff's Guide to Intuitive Healing*. New York: Times Books.

Randi, James. 1982. *Flim-Flam!* Buffalo, N.Y.: Prometheus Books.

The singular death of Mr. Matteson. 1884. *Buffalo Express*, 18 March.

Stearn, Jess. 1968. *Edgar Cayce—The Sleeping Prophet*. New York: Bantam Books.

Turner, Gladys Davis, and Mae Gimbert St. Clair. 1976. *Individual Reference File of Extracts from the Edgar Cayce Readings* [note on copyright page]. Virginia Beach, Va.: Edgar Cayce Foundation.

Whitcomb, Nellie. 1923. Advertising flyer for "Mrs. J. H. R. Matteson's Psychic Clairvoyant Remedies," containing a biographical sketch of Mrs. Matteson. Copy in University Archives, University at Buffalo.

25

Alien Abductions as Sleep-Related Phenomena

In his book, *The Communion Letters* (1997), self-claimed alien abductee Whitley Strieber, assisted by his wife Ann, offers a selection of letters Strieber has received in response to his various alien-abduction books, particularly the best-selling *Communion: A True Story* (1987). A careful analysis of these letters is illuminating.

The 67 narratives that constitute *The Communion Letters* represent, the Striebers claim, what "could conceivably be the first true communication from another world that has ever been recorded." Selected from nearly 200,000 letters, those in the published collection, they assert, "will put certain shibboleths to rest forever": namely, that the phenomenon is limited to a few people, that people are always alone when abducted, that the events are recalled only under hypnosis, and that the abductees are attention-seekers (Strieber and Strieber 1997, 3–4). (For a discussion of the role of hypnosis in claims of alien abduction, see Nickell 1997.)

Be that as it may, the accounts are really surprising for the prevalence of simple, well-understood, sleep-associated phenomena related therein. Most, for example, include one or more experiences that can easily be attributed to some type of dream.

The dominant phenomenon in the accounts—albeit one that is little known to the public—is clearly the common "waking dream." This occurs when the subject is in the twilight state between waking and sleeping, and combines features of both. Such dreams typically include perception of bright lights or other bizarre imagery, such as apparitions of strange creatures. Auditory hallucinations are also possible. Waking dreams are termed *hypnagogic* hallucinations if the subject is going to sleep, or *hypnopompic* if he or she is awakening (Drever 1971, 125). Frequently the latter is accompanied by what is known as *sleep paralysis*, an inability to move caused by the body's remaining in the sleep mode.

In the Middle Ages, waking dreams were often responsible for reports of demons (incubuses and succubuses) which, due to sleep paralysis, sometimes seemed to be sitting on the percipient's chest or lying atop his or her body. In other eras, waking dreams have been common sources of reports of "ghosts," "angels," and other imagined entities (Nickell 1995, 41, 46, 55, 59, 117, 131, 157, 209, 214, 268, 278). Now, as the collection of letters to the Striebers demonstrates, these experiences are producing reports of "aliens" and related imagery. Some 42 of the 67 narratives include one or more apparent waking dreams.

For example, one man wrote:

> I'd wake up and my heart would be pounding as if I was frightened. I'd also see two white lights, one slightly higher than the other, flying or floating across my room in a descending motion toward the floor. . . . I would have

> what I called a 'dream,' although I felt that I was totally awake because I could move my eyes. My body would be completely paralyzed. I couldn't yell or scream, but wanted to" [Strieber and Strieber 1997, 87].

Another man reported:

> At night, after my parents would put me to bed, I'd often see small, very white round faces with huge black eyes staring in at me from outside my bedroom window. Sometimes it was only one, but often it was several. . . . I saw them several nights a week almost into my teens [37].

Still another man wrote: "When I was twenty-three I woke up one night to find a little gray man on the other side of my room. He looked about four feet tall and had very large orange cat eyes. I later learned that this was my 'guardian'" (135).

A woman reported:

> One night while soundly asleep and in a dream state, the dream was suddenly interrupted by a loud noise and the appearance of a stark white face and head, which faded into and out of focus several times, directly in front of me. Although I felt I was fully conscious, my eyes were closed. . . . I convinced myself that I'd experienced some unusual form of nightmare [58].

As these accounts show, some of the "abductees" do not report paralysis, although others describe that effect without imagery. For example, one wrote: "I woke up into one of the strangest experiences of my life. I was awake, could feel and could smell and think and reason, but I could not see. . . . I experimented with every part of my body to see if I could move; I couldn't. There was a flashlight a few inches from my head, but I couldn't make my arm respond to my mental commands" (40).

Sleep paralysis accompanying a waking dream may well be a major factor in convincing some "abductees" that they have been examined by aliens. Consider this woman's account:

> I often found myself being awakened in the deepest night by a feeling of someone touching me: pushing my stomach; poking my arms and legs; touching my head and neck; what felt like a breast exam and a heaviness across my chest, and someone holding my feet. This seemed to go on for three nights. On the last night, I vaguely saw, in my efficien[cy] apartment, a "little man" running to and around my refrigerator. My door was always locked, as were the two windows.
>
> Then one night I woke up to find myself in a strange room, *strapped to a table*, with my feet up. I felt that my lower half was undressed. . . . On another later night, I woke up strapped to a table in a reclining position [250–51, emphasis added].

Another phenomenon reported in *The Communion Letters* is the out-of-body experience (OBE). This may be associated with a waking dream, as in this woman's account:

> When I was nineteen I had my first OBE. . . . I should say here that, to my knowledge, all my hundreds of OBEs throughout the years have been conscious ones, meaning that they've all occurred in the state just before sleep, where I am fully conscious and aware of the paralysis, the vibrations that occur, and of the actual separation. . . . On the night of March 15, 1989, I went to bed and fell asleep

> normally. Sometime during the night I awakened to find myself softly bumping against the ceiling, already separated from the physical. . . . I felt myself being turned around. I "saw" a being standing in the middle of the open room, approximately fifteen feet away. A telepathic voice asked if I was afraid.

The woman then goes on to describe a stereotypical alien (73–74).

Such "telepathic" voices—which are often part of a waking dream—are, of course, the person's own. Even abduction guru David Jacobs admits that reports of telepathic communication with aliens may be nothing more than confabulation (the tendency of ordinary people to confuse fact with fantasy [Baker and Nickell 1992, 217]). Says Jacobs, "Abductees sometimes slip into a 'channeling' mode—in which the abductee 'hears' messages from his own mind and thinks they are coming from outside sources—and the researcher fails to catch it" (Jacobs 1998, 56).

No fewer than 18 letters in the Strieber collection describe one or more OBEs, or such related phenomena as "astral travel" or floating or flying dreams. The relationship between OBEs and sleep paralysis is demonstrated by a percipient who had "the strangest type of dream" up to three times a week. He would awaken to hear crackling noises followed by a loud boom, "at which point," he says, "I would immediately go into paralysis. Then I would slowly begin to float toward my ceiling, unable to move a limb" (Strieber and Strieber 1997, 130).

In a few instances the "abductee" is not in bed when the (apparent) waking dream occurs. He or she may be watching television, riding in a car, or—as in the case of one woman—sitting with her child in a rocking chair. "We must have rocked for twenty minutes, and I was actually becoming drowsy. My eyes were closed. Then an odd thing happened: I got a vision of three 'grays' standing in front of the rocking chair. It was as if I could see through my eyelids" (17). The salient point is that the waking dream may occur virtually anywhere, as long as the person is in the state between waking and sleeping.

In fact, the subject may have experiences similar to those in waking dreams when he or she is simply exhausted; that is, suffering from mental or physical fatigue (Baker 1992, 273). Such might be the explanation for eight of the reports, like that of one woman who told the Striebers:

> I was going home from work [i.e., presumably tired], and in the middle of the Seventh Avenue subway rush hour crowd I saw a little man about four feet tall. He had a huge head, but it was the quality of his skin that first caught my attention. It didn't look like human skin, but more like plastic or rubber. I knew he wasn't human. I tried to follow him with my eyes, but he quickly got lost in the crowd. No one else seemed to notice. This disturbed me; *I thought I was seeing things* [emphasis added].

This person also had "recurrent dreams" of "spaceships hovering over the Hudson River and the Palisades. These dreams were always very vivid and powerful" (207).

Other accounts in *The Communion Letters* clearly indicate ordinary dreaming, nightmares, "lucid" dreams (vivid dreams that occur when one is fully asleep), and the like—in all, reports by some 22 letter writers. At least four reports almost certainly involved somnambulism (walking or performing other activities while asleep). The letters also reported "near-death experiences" (two writers) and hypnosis (another two instances). A majority of the narratives contain more than one phenomenon, but in all at least 59 of the 67 letters consist of one or more instances of probable sleep-related phenomena such as those discussed thus far. (In addition, there were such reported conditions as migraines, panic attacks, posttraumatic stress disorder, and even schizophrenia—one example of each. As many as eight people had a number of the traits associated with what is termed "fantasy proneness.")

Lest it be thought that the eight remaining letters are reports of genuine abductions, I consider three to be extremely doubtful, raising more questions than they answer and even containing internal inconsistencies or outright contradictions. Of the other five, two are reports of nothing more than unexplained knocking sounds and three consist merely of rather typical UFO sightings (two possibly weather balloons), with one writer specifically stating, "I do not believe that an abductee experience is in my recent history" (180).

Strieber's correspondents have, of course, read his books, *Communion*, *Transformation*, and *Breakthrough*, and they clearly have been influenced by them. Indeed, one writer's experience with "the visitors"—an alleged abduction—"happened the night after I finished your last book, *Breakthrough*" (Strieber and Strieber 1997, 144)! Another, who has "had plenty of UFO experiences," wrote: "I couldn't get the picture of the being on the *Communion* cover out of my head" (134–35). A woman stated: "When I saw the cover of *Communion* I felt compelled to buy it. When I began to read it, I felt nauseated, burst into tears, was shaking, and was elated. Most books don't elicit this reaction in me as I read the first few chapters" (148). A policeman wrote: "Frankly the books scare the hell out of me. I did not sleep well for weeks following *Communion*. I again feel very restless after reading *Breakthrough*. I cannot explain this. Tell me I am imagining things" (122). Obviously, such correspondents are quite impressionable, to say the least.

Many who wrote did so in response to similar events reported by Strieber. Significantly, Strieber's own abduction claims began with his having a waking dream! According to psychologist Robert A. Baker (1987, 157):

> In Strieber's *Communion* is a classic, textbook description of a hypnopompic hallucination, complete with the

> awakening from a sound sleep, the strong sense of reality and of being awake, the paralysis (due to the fact that the body's neural circuits keep our muscles relaxed and help preserve our sleep), and the encounter with strange beings. Following the encounter, instead of jumping out of bed and going in search of the strangers he has seen, Strieber typically goes back to sleep. He even reports that the burglar alarm was still working—proof again that the intruders were mental rather than physical. Strieber also reports an occasion when he awakes and believes that the roof of his house is on fire and that the aliens are threatening his family. Yet his only response to this was to go peacefully back to sleep. Again, clear evidence of a hypnopompic dream. Strieber, of course, is convinced of the reality of these experiences. This too is expected. If he was not convinced of their reality, then the experience would not be hypnopompic or hallucinatory.

Why some people's waking dreams relate to extraterrestrials and others to different entities depends on the person's expectations, which in turn are influenced by various cultural and psychological factors. Thus, given different contexts, a waking dream involving a shadowy image and sleep paralysis may be variously reported: someone sleeping in a "haunted" manor house describes a ghostly figure and is "paralyzed with fear"; another, undergoing a religious transformation, perceives an angel and is "transfixed with awe"; yet another, having read *Communion*, sees an extraterrestrial being and feels "strapped to an examining table."

Many of the communicants in *The Communion Letters* even show a willingness to reinterpret their original experiences in light of what they have since read in Strieber's books. This transformational tendency seems quite strong. One woman, for example, who had "imaginary playmates" as a child in the 1940s, now reports to Strieber: "The beings that I saw looked like the ones in your book" (Strieber and Strieber 1997, 93). Another, who saw an entity during an obvious waking dream, reported that her first reaction after reading *Communion* "was to wonder if, in fact, what I recalled was all that had taken place the night of my experience" (119). Still another, a man who would sometimes "wake up with little gray people around me," admitted: "I never associated them with UFOs. As soon as I'd open my eyes, they'd all run away, right through the walls!" (134). Now that he has read *Communion,* he believes he was "manipulated" into buying it. This same person also had a "memory" that "came in the form of a vivid dream" and that involved himself, Strieber, and the aliens. "When I awoke," he reported, "I felt as if you had been looking at me intently" (136). In *The Threat*, David Jacobs even tries to convince his readers that they should revise their perceptions of their experiences. He suggests that their "ghost" or "guardian angel" experiences should be considered possible alien encounters, and that they may therefore be "unaware abductees" (Jacobs 1998, 120).

It is distressing that such simple phenomena as waking dreams, sleep paralysis, and out-of-body hallucinations can be transformed into "close encounters." The mechanism is what psychologists call *contagion*—the spreading of an idea, behavior, or belief from person to person by means of suggestion (Baker and Nickell 1992, 101). Examples of contagion are the Salem witch hysteria of 1692–1693; the spiritualist craze of the nineteenth century; the UFO furor that began in 1947; and, of course, its sequel, today's alien-encounter delusion, the dissemination of which is aided by the mass media.

Perhaps we should not be surprised that those who are hyping belief in extraterrestrial abductions ignore or underestimate the psychological factors. Strieber, for example, is a fiction writer, and Budd Hopkins, who helped boost public interest with his 1981 book *Missing Time*, is an artist. However, one would think that history professor David

Jacobs would profit from the mistakes of the past and not help repeat them. Even more curious is the involvement of clinical psychologist Edith Fiore (1989) and psychiatrist John Mack (1994). Both confess, though, that they are less interested in the truth or falsity of a given claim than in what the individual *believes* happened, and the resulting significance to therapy and, in the case of Mack, to "the larger culture" (Mack 1994, 382; see also Fiore 1989, 333–34; Jacobs 1998, 48–55).

All of these abduction promoters have books to offer. Let the buyer beware.

REFERENCES

Baker, Robert A. 1987. The aliens among us: Hypnotic regression revisited. *Skeptical Inquirer* 12, no. 2: 147–62.

———. 1992. *Hidden Memories: Voices and Visions from Within*. Buffalo, N.Y.: Prometheus Books.

Baker, Robert A., and Joe Nickell. 1992. *Missing Pieces: How to Investigate Ghosts, UFOs, Psychics, and Other Mysteries*. Amherst, N.Y.: Prometheus Books.

Drever, James. 1971. *A Dictionary of Psychology*. Baltimore: Penguin Books.

Fiore, Edith. 1989. *Encounters: A Psychologist Reveals Case Studies of Abductions by Extraterrestrials*. New York: Doubleday.

Jacobs, David. 1998. *The Threat*. New York: Simon & Schuster.

Mack, John. 1994. *Abduction: Human Encounters with Aliens*. New York: Scribners.

Nickell, Joe. 1995. *Entities: Angels, Spirits, Demons, and Other Alien Beings*. Amherst, N.Y.: Prometheus Books.

———. 1997. A study of fantasy proneness in the thirteen cases of alleged encounters in John Mack's *Abduction*. In *The UFO Invasion*, edited by Kendrick Frazier, Barry Karr, and Joe Nickell. Amherst, N.Y.: Prometheus Books.

Strieber, Whitley. 1987. *Communion: A True Story*. New York: William Morrow.

Strieber, Whitley, and Ann Strieber. 1997. *The Communion Letters*. New York: HarperPrism.

26

"Visitations"

After-Death Contacts

Those who suffer the loss of a loved one may experience such anguish and emptiness that they are unable to let go, and they may come to believe that they have had some contact with the deceased. "It's commonly reported that the deceased person has communicated in some way," says Judith Skretny, vice-president of the Life Transitions Center, "either by giving a sign or causing things to happen with no rational explanation." She adds, "It's equally common for people to wake in the middle of the night, lying in bed, or even to walk into a room and think they see their husband or child" (quoted in Voell 2001). These experiences are sometimes called *visitations* (Voell 2001), and they include deathbed visitations (Wills-Brandon 2000).

During more than 30 years of paranormal investigation, I have encountered countless claims of such direct contacts (as opposed to those supposedly made through spiritualist mediums [Nickell 2001a; 2001b]).

I have also occasionally been interviewed on the subject—most recently in response to some books promoting contact claims (Voell 2001). Here is a look at the evidence regarding purported signs, dream contacts, apparitions, and deathbed visions.

"Signs"

In her co-authored book *Childlight: How Children Reach Out to Their Parents from the Beyond*, Donna Theisen relates a personal contact she believes she received from her only son, Michael, who was killed in an auto accident a month before. She was browsing in a gift shop when she noticed a display of dollhouse furnishings. Nearby, on a small hutch, were a pair of tiny cups that were touching, one bearing the name "Michael," the other the words "I love you, Mom." Although at the time a "strange, warm feeling" came over her, she was later to wonder: "Was I merely finding what I so desperately wanted to see? Was I making mystical connections out of ordinary circumstances?"

However, the fact that those two cups were displayed together, out of the dozens of others sold there, convinced Theisen that the incident "defied the odds." Soon she "began looking for more strange occurrences" so as to confirm that the cups incident was "a real sign." Her book chronicles them and the experiences of other grieving parents (40 of 41 of them mothers). One, whose son was killed by a train, was wondering whether to give her son's friend some of his baseball equipment when she heard a train whistle blow; she accepted this as an affirmation. Others received signs in the form of a rainbow, television and telephone glitches, the arrival and sudden departure of pigeons, a moved angel doll, and other occurrences (Theisen and Matera 2001).

To explain such signs or "meaningful coincidences" (conjunctions of events that seem imbued with mystical significance), psychologist Carl Jung (1960) theorized that—in addition to the usual cause-and-effect relationship of events—there is an "acausal connecting principle." He termed this *synchronicity*. However, in *The Psychology of Super-*

stition, Gustav Jahoda (1970, 118) suggests there may often be causal links of which we are simply unaware.

Even in instances in which there may in fact be no latent causal connections, other factors could apply. One is the problem of overestimating how rare an occurrence really is. Nobel Prize-winning physicist Luis Alvarez (1965) told how, while reading a newspaper, he came across a phrase that triggered certain associations and left him thinking of a long-forgotten youthful acquaintance; just minutes afterward, he came across that person's obituary. On reflection, Alvarez assessed the factors involved, worked out a formula to determine the unlikeliness of such an event, and concluded that 3,000 similar experiences could be expected each year in the United States, or approximately 10 per day. Synchronous events involving family and friends would be proportionately more common.

A related problem is what psychologist Ruma Falk (1981–1982) terms "a selection fallacy" that occurs with anecdotal events as contrasted with scientifically selected ones. As he explains:

> Instead of starting by drawing a random sample and then testing for the occurrence of a rare event, we select rare events that happened and find ourselves marveling at their nonrandomness. This is like the archer who first shoots an arrow and then draws the target circle around it.

Some occurrences that are interpreted as signs probably have mundane explanations. Although unexplained, they are not unexplainable. For example, the mother of a severely handicapped little boy reported that on the morning of his funeral, she awoke to see a small, glowing red light on the dresser where his baby monitor had been. It was in fact a tiny lantern on her keychain. "It had never been turned on before," she said. "In fact, I didn't even know it worked! The moment I touched the light, it went out." This happened for several subsequent mornings (Theisen and Matera 2001, 192). How do we explain such a mystery? One possibility is that the light was not turned on at all, but only appeared to be lit as sunlight reflected off its red cover; when it was picked up, the illusion was dispelled.

Photographic "signs," which are also becoming common, may be easily explained. I recall a Massachusetts woman approaching me after a lecture to show me some "ghost" photographs. I immediately recognized the white shapes in the pictures as resulting from the camera's flash bouncing off the stray wrist strap—a phenomenon I had previously investigated and replicated (Nickell 1996). In fact, in one snapshot, the strap's adjustment slide was even recognizable, silhouetted in white. But the lady would not hear my explanation; instead, she took back the pictures and stated defiantly that her father had recently died and had been communicating with the family in a variety of strange ways.

In addition to numerous glitches caused by camera, film, and other factors, photographs may also exhibit *simulacra*, random shapes that are interpreted, like inkblots, as recognizable figures (such as a profile of Jesus seen in the foliage of a vine-covered tree [Nickell 1993, 34–41]). These can easily become visitation signs, as in the case of a photo snapped from a moving vehicle at the site of a young man's auto-accident death. "When this photo was developed," the victim's mother wrote, "the tree branches formed a startling figure that looked just like Greg wearing his hat. In addition, there appeared to be an angel looking out toward the road." She added, "we all viewed this photo as more evidence of Greg's ongoing existence" (Theisen and Matera 2001, 47).

Dream Contacts

A significant number of after-death "contacts" come from dreams. They have been associated with the supernatural since very ancient times, and attempts to interpret them are recorded in a papyrus of 1350 B.C.E. in the British Museum (Wortman and Loftus 1981, 380). Now New Age writers like Theisen and Matera (2001) are increasingly chronicling instances of people having dreams about their departed loved ones.

It has been estimated that the average person will have approximately 150,000 dreams by the time he or she reaches the age of 70. Although most are forgotten, the more dramatic and interesting ones are often remembered and talked about (Wortman and Loftus 1981, 380). However, people's reports of their dreams may be undependable, because of the effects of memory distortion, ego, superstition, and other factors.

Even an ordinary dream can be especially powerful when it involves after-death content, and there are types of dreams that can be extremely vivid and seemingly real. They include lucid dreams in which the dreamer is able to *direct* the dreaming, "something like waking up in your dreams" (Blackmore 1991a).

A powerful source of visitation reports is the so-called waking dream, which occurs in the twilight between wakefulness and sleep and combines features of both. Actually an hallucination—called *hypnagogic* if the subject is going to sleep or *hypnopompic* if he or she is awakening—it typically includes bizarre imagery such as apparitions of ghosts, angels, aliens, or other imagined entities. The content, according to psychologist Robert A. Baker (1990), "may be related to the dreamer's current concerns."

For example, here is an account I obtained in 1998 from a Buffalo, New York, woman:

> My father had passed away and I was taking care of my sick mother. I went to lay down to rest. I don't remember if I actually fell asleep or if I was awake, but I saw the upper part of my father and he said, "Mary Ellen, you're doing a good job!" When I said "Dad," he went away.

It would be correct to say that this describes a rather common hypnagogic event; nevertheless, this does not do justice to the person who experienced it. For her, I think, it represented a final goodbye from her father, and therefore a form of closure, and also provided welcome reassurance during a period of difficulty.

Sometimes, a waking dream is accompanied by what is termed *sleep paralysis*, an inability to move caused by the body remaining in the sleep mode. Consider this account (Wills-Brandon 2000, 228–29):

> My sister said she was abruptly awakened from a very deep sleep. When she woke up, she said her body felt frozen and she couldn't open her eyes. Suddenly she felt a presence in the room and knew it was Mother. She felt her standing at the foot of the bed.

By their nature, waking dreams seem so real that the experiencer typically insists that he or she was not dreaming. One woman, who "hardly slept" after her daughter's suicide, saw the daughter, late at night, standing at the end of a long hallway, smiling sadly and then walking away into a brilliant light. "At first I thought I was hallucinating," the mother said. "But after a new round of tears, I realized that I was wide awake and I had indeed seen Wendy" (Theisen and Matera 2001, 130). Another, describing a friend's "visitation" experience of her deceased mother-in-law, said, "At first my friend thought she was dreaming but quickly realized she was wide awake" (Wills-Brandon 2000, 60)—a confusion typical of a waking dream.

Apparitions

Some visitations are reported as quite undreamlike, in that they occur during normal daily activity. However, my own investigatory experience, as well as other research data, demonstrates that apparitions are most apt to be perceived during daydreams or other altered states of consciousness. Many occur, for example, while the percipient is in a relaxed state, or concentrating on some activity like reading, or performing routine work. In some instances the person may simply be tired, as from a long day's work. Under such conditions, particularly in the case of imaginative individuals, a mental image might be superimposed upon the visual scene to create a "sighting" (Nickell 2001a, 291–92).

Also, as indicated earlier, faulty recall, bias, and other factors can betray even the most credible and sincere witness. Consider, for instance, an anecdotal case provided by Sir Edmund Hornby, a Shanghai jurist. He related how, years earlier, he was awakened one night by a newspaperman who had arrived belatedly to get the customary written judgment for the following day's edition. The man—looking "deadly pale"—would not be put off, and sat on the jurist's bed. Eventually Judge Hornby provided a verbal summary, which the man took down in his pocket notebook. After the visitor left, the judge related the incident to Lady Hornby. The following day the judge learned that the reporter had died during the night; more importantly, he discovered that the man's wife and servants were certain he had not left the house; yet with his body was discovered a notebook containing a summary of Hornby's judgment!

This apparent proof of a visitation was reported by psychical researchers. However, the tale soon succumbed to investigation. As it turned out, the reporter did not die at the time reported (about 1:00 A.M.) but much later—between 8:00 and 9:00 in the morning. Furthermore, the judge could not have told his wife about the events at the time, because he was then between marriages. Finally, although the story depends on a certain judgment that was to be delivered the following day, no such judgment was recorded (Hansel 1966, 186–89).

When confronted with this evidence of error, Judge Hornby admitted: "My vision must have followed the death (some three months) instead of synchronizing with it . . . . " Bewildered by what had happened, he added: "If I had not believed, as I still believe, that every word of it [the story] was accurate, and that my memory was to be relied on, I should not have ever told it as a personal experience." No doubt many other accounts of alleged visitations involve such *confabulation*—a term psychologists use to refer to the confusing of fact with fiction. Unable to retrieve something from memory, the confabulating person (perhaps inadvertently) manufactures something that is seemingly appropriate to replace it. "Thus," explain Wortman and Loftus (1981, 204), "the man asked to remember his sixth birthday combines his recollections of several childhood parties and invents the missing details."

Tales such as that related by Judge Hornby represent alleged "moment-of-death visitations" (Finucane 1984, 195). In that story the reporter had allegedly died at approximately the same time ("about twenty minutes past one") that he appeared as an apparition to Judge Hornby—although, as we have seen, the death actually occurred several hours later. This case should serve as a cautionary example regarding other such accounts, which are obviously intended to validate superstitious beliefs.

Deathbed Visions

Another type of alleged visitation comes in the form of deathbed visions. According to Brad Steiger (real name Eugene E. Olson), who endlessly cranks out books promoting paranormal claims, "The phenomenon of deathbed visions is as old as humankind, and such visitations of angels, light beings, previously deceased personalities and holy figures manifesting to those about to cross over to the Other Side have been recorded throughout all of human history." Steiger (2000) goes on to praise writer and family grief counselor Carla Wills-Brandon for her "inspirational book," *One Last Hug Before I Go: The Mystery and Meaning of Deathbed Visions* (2000).

Like others before her (e.g., Kübler-Ross 1973), Wills-Brandon promotes deathbed visions (DBVs) largely through anecdotal accounts —which, as we have seen, are untrustworthy. She asserts that "the scientific community" has great difficulty explaining a type of DBV in which the dying supposedly see people they believe are among the living but who have actually died. She cites an old case involving a Frenchman who died in Venezuela in 1894. His nephew(who had not been present at the scene) reported:

> Just before his death, and while surrounded by all of his family, he had a prolonged delirium, during which he called out the names of certain friends left in France. . . .
>
> Although struck by this incident, nobody attached any extraordinary importance to these words at the time they were uttered, but they acquired later an exceptional importance when the family found, on their return to Paris, the funeral invitation cards of the persons named by my uncle before his death, and who had died before him.

Unfortunately, when we hear two other accounts of these events, we find there is less to this story than meets the ear. A version given by one of the man's two children says nothing of his being delirious, implying otherwise by stating that "he told us of having seen some persons in heaven and of having spoken to them at some length." But she had been quite young at the time and referred the inquirer to her brother. His account—the most trustworthy of the three, because it is a firsthand narrative by a mature informant—lacks the multiple names, and the corresponding funeral cards, as well as other elements, thus indicating that the story has been much improved in the retellings. The son wrote:

> Concerning what you ask me with regard to the death of my father, which occurred a good many years ago, I recall that a few moments before his death my father called the name of one of his old companions—M. Etcheverry—with whom he had not kept up any connexion, even by correspondence, for

> a long time past, crying out, "Ah! you too," or some similar phrase. It was only on returning home to Paris that we found the funeral card of this gentleman.

He added, "Perhaps my father may have mentioned other names as well, but I do not remember."

It is hardly surprising that a man's thoughts should, at the close of life, turn to an old friend, or that—having long been out of touch—he should have thought the friend already dead. (The individual reporting the case conceded that there was no *certainty* that the friend had died before the vision occurred.) As the most trustworthy account is the least elaborate, lacking even the vision-of-heaven motif, it seems not a corroboration of the nephew's hearsay accounts (Barrett 1926, 22–24) but rather evidence of confabulation at work.

In their book *The Afterlife*, Jenny Randles and Peter Hough (1993, 98–99) tell of a dying man who had lapsed into a coma:

> Then the patient became wonderfully alert, as some people do very near the end. He looked to one side, staring into vacant space. As time went by it was clear he could see someone there whom nobody else in the room could see. Suddenly, his face lit up like a beacon. He was staring and smiling at what was clearly a long-lost friend, his eyes so full of love and serenity that it was hard for those around him to not be overcome by tears.
>
> Sheila [his nurse] says: "There was no mistake. Someone had come for him at the last to show him the way."

But how did the nurse know it was "a long-lost friend" and not, say, Jesus or an angel? Indeed, how did she know he saw "someone" at all, rather than some*thing*—perhaps an entrancing view of heaven? The way the nurse makes such assertions—emphasized with words like "clear" and "no mistake"—suggests that she is speaking more of faith than of fact, and her belief is accepted and reported uncritically by Randles and Hough. In fact, the tale contains no evidence of a visitation at all.

Instead, it appears to represent what is termed a *near-death experience* (NDE), in which a person typically "comes back" from a state close to death with a story of an otherworldly visit, sometimes involving an out-of-body experience, travel down a dark tunnel, and an encounter with beings of light who help him or her decide whether to cross over.

Susan Blackmore (1991b) describes the NDE as "an essentially physiological event" prompted by lack of oxygen, the structure of the brain's visual cortex, and other factors. She recognizes that the experiences are hallucinations—albeit, seemingly, exceedingly real. Also, she points out that one does not actually have to be near death to have such an experience: "Many very similar experiences are recorded of people who have taken certain drugs, were extremely tired, or, occasionally, were just carrying on their ordinary activities."

Many of the DBVs reported by Wills-Brandon (2000) and others are similar to NDEs and are probably hallucinations produced by the dying brain. Some of the effects are similar because people share similar brain physiology. For example, the "tunnel" effect "probably lies in the structure of the visual cortex" (Blackmore 1991b, 39–40). Other effects are probably psychological and cultural. Wills-Brandon (2000, 115) concedes: "I agree that when the dying are passing, they are visited by those who will comfort them during their travel to the other side. For a dying Christian, that might mean Jesus; a Buddhist may see Buddha. For others, an angel, a beautiful woman or Druid priest would bring more comfort." But she rationalizes, "If I'm following a particular philosophy of religion, wouldn't it make sense for me to be visited at the moment of my death by an otherworldly escort who is familiar with my belief system?" Perhaps, but of course the simpler explanation is that people see what their expectations prompt them to see.

And that, in a nutshell, is the problem with the anecdotal evidence for so-called visitations. The experiencer's will to believe may override any temptation to critically examine the occurrences. Some proponents of after-death contact adopt an end-justifies-the-means attitude. One (quoted in Voell 2001) states: "Whether any of the connections or feelings or appearances are true or not, I've finally figured out it doesn't make a damn bit of difference. If it has any part in healing, who cares?" The answer is that, first of all, people who value truth care. Although magical thinking may be comforting in the short term, over time estrangement from rationality can have consequences, both on individuals, who may suffer from a lack of closure, and societies, which may slide into ignorance and superstition. That potential peril is why Carl Sagan (1996) referred to science as "A Candle in the Dark."

REFERENCES

Alvarez, Luis W. 1965. A pseudo experience in parapsychology. Letter in *Science* 148: 1541.

Baker, Robert A. 1990. *They Call It Hypnosis*. Buffalo: Prometheus Books.

Barrett, Sir William. [1926] 1986. *Death-Bed Visions: The Psychical Experiences of the Dying*. Reprinted Wellingborough, England: The Aquarian Press.

Blackmore, Susan. 1991a. Lucid dreaming: Awake in your sleep? *Skeptical Inquirer* 15, no. 4 (Summer): 362–70.

———. 1991b. Near-death experiences: In or out of the body? *Skeptical Inquirer* 16, no. 1 (Fall): 34–45.

Falk, Ruma. 1981–1982. On coincidences. *Skeptical Inquirer* 6, no. 2 (Winter): 24–25.

Finucane, R. C. 1984. *Appearances of the Dead: A Cultural History of Ghosts*. Buffalo, N.Y.: Prometheus Books.

Hansel, C. E. M. 1966. *ESP: A Scientific Evaluation*. New York: Scribner's.

Jahoda, Gustav. 1970. *The Psychology of Superstition*. Baltimore: Penguin.

Jung, C. G. 1960. Synchronicity: An acausal connecting principle. In *The Collected Works of C. G. Jung*, Bollingen Series, no. 20, edited by Sir Herbert Read et al., 418–519. New York: Pantheon.

Kübler-Ross, Elizabeth. 1973. *On Death and Dying*. London: Tavistock.

Nickell, Joe. 1993. *Looking for a Miracle*. Amherst, N.Y.: Prometheus Books.

———. 1996. Investigative files: Ghostly photos. *Skeptical Inquirer* 20, no. 4 (July/August): 13–14.

———. 2001a. *Real-Life X-Files: Investigating the Paranormal.* Lexington, Ky.: University Press of Kentucky.

———. 2001b. John Edward: Hustling the bereaved. *Skeptical Inquirer* 25, no. 6 (November/December): 19–22.

Randles, Jenny, and Peter Hough. [1993] 1995. *The Afterlife: An Investigation into Life after Death.* Reprinted London: BCA.

Sagan, Carl. 1996. *The Demon-Haunted World: Science as a Candle in the Dark.* New York: Random House.

Steiger, Brad. 2000. Promotional blurb. In *One Last Hug Before I Go: The Mystery and Meaning of Deathbed Visions,* by Carla Wills-Brandon. Deerfield Beach, Fla.: Health Communications, Inc.

Theisen, Donna, and Dary Matera. 2001. *Childlight: How Children Reach Out to Their Parents from the Beyond.* Far Hills, N.J.: New Horizon Press.

Voell, Paula. 2001. Visitations. *Buffalo News* (Buffalo, N.Y.), 27 May.

Wills-Brandon, Carla. 2000. *One Last Hug Before I Go: The Mystery and Meaning of Deathbed Visions.* Deerfield Beach, Fla.: Health Communications, Inc.

Wortman, Camille B., and Elizabeth F. Loftus. 1981. *Psychology.* New York: Knopf.

27

The Sacred Cloth of Oviedo

Although science has established the Shroud of Turin (see chapter 22) as a fourteenth-century forgery—rendered in tempera paint by a confessed forger and radiocarbon-dated to the time of the forger's confession (Nickell 1998, McCrone 1996)—the propaganda offensive to convince the public otherwise continues. As part of the strategy, shroud proponents are now ballyhooing another cloth, a supposed companion burial wrapping, that they claim militates in favor of the shroud's authenticity.

"Companion Relic"

At issue is the Oviedo Cloth, an 84 × 53 cm. piece of linen, stained with supposed blood, that some believe is the *sudarium* or "napkin" that covered the face of Jesus in the tomb. As described in the New Testament (John 20:7) it was "about his head." Such a cloth was used in ancient Jewish burial practice to cover the face of the deceased (Nickell 1998, 33–34).

One reason for the interest in the Oviedo Cloth among Shroud of Turin advocates is to counter the devastating radiocarbon evidence. Three laboratories used sophisticated C-14 dating technology to test a piece of shroud cloth, and the resulting age span was found to be circa C.E. 1260–1390. In response, advocates hope to tie the shroud to the Oviedo Cloth because, allegedly, "the history of the sudarium is undisputed" and it "was a revered relic preserved from the days of the crucifixion" (Anderson 2000).

Unfortunately for the shroudologists, however, the provenance (or historical record) of the Oviedo Cloth, currently located in the Cathedral of Oviedo in northern Spain, is not nearly so definitive. Indeed, even most pro-authenticity sources admit that it cannot really be established as dating from earlier than about the eighth century (Whanger and Whanger 1998, 56), and the earliest supposed documentary evidence is from the eleventh century. According to Mark Guscin in *The Oviedo Cloth* (1998, 17), "The key date in the history of the sudarium is 14 March 1075." At that time, an oak chest in which the cloth was kept was reportedly opened by King Alfonso VI and others, including the famed knight El Cid; this is recorded in a document stating that the chest had long reposed in the church. Unfortunately, the original document has been lost, and only a thirteenth-century "copy" remains in the cathedral archives (Guscin 1998, 17).

An account of the cloth was penned in the twelfth century by a bishop of Oviedo named Pelayo, who claimed that the sudarium had been kept in Jerusalem from the time it was discovered in the tomb until the seventh century, when Christians fleeing the Persian invasion took it to Spain. However, relic mongers typically fabricated stories about their bogus productions, and there were many allegedly genuine *sudaria*, just as there were numerous "true shrouds"—at least 43 in medieval Europe alone (Humber 1978, 78). Furthermore, there is not the slightest hint in the Christian gospels, or anywhere else in the New Testament, that the burial wrappings of Jesus were actually preserved. Later, of course, certain apocryphal texts claimed otherwise. One fourth-century account mentioned a tradition that Peter had kept the sudarium, but what subsequently became of it was unknown (Wilson 1979, 92–95).

Those who would try to link the questionable Oviedo sudarium to the Turin "shroud"—and vice-versa, in the hopes of mutual authentication—face a problem: The sudarium lacks an image like that on the shroud. Had such a cloth indeed covered the face of Jesus, "this would have prevented the image from being formed on the shroud, and it would presumably have caused it to be formed on the sudarium" instead (Guscin 1998, 33, 34). Proponents now postulate that the sudarium was used only *temporarily*, in the period after crucifixion and before burial, and that it was put aside before the body was wrapped.

However clever this rationalization, John's gospel states that Jesus was buried "as the manner of the Jews is to bury" (19:40), and the use of a kerchief to cover the face in burial is specifically mentioned in the Jewish Mishnah. Also, with regard to the burial of Lazarus (John 11:44), who was "bound hand and foot with graveclothes," we are told that "his face was bound about with a napkin."

Undaunted, shroud and sudarium advocates have joined forces and are now making the kind of outrageous and pseudoscientific claims that used to be made for the shroud alone, declaring that "blood" and pollen evidence link the two cloths. Unfortunately, the new claims come from many of the same dubious and discredited sources as before.

"Blood" Stains

At an international congress in Oviedo in October 1994, papers were presented focusing on the latest "investigations" of the supposed sudarium. One claimant was a Dr. Pierluigi Baima-Bollone, who purported to have established that the "blood" stains on the cloth were not only human blood but were also blood of type AB—"the same group," according to Guscin (1998, 56), "as the blood on the shroud."

Actually, the assertion that the shroud has type AB blood comes from the same source, and Bollone's claims are baloney. Even one of the shroud's most committed defenders, Ian Wilson (1998, 89), merely remarks in passing that Bollone "claimed to" have made such a determination. A zealous shroud partisan and chairman of a shroud center, Bollone is a professor of legal medicine.

In contrast, internationally known forensic serologists, employing the standard scientific tests used in crime laboratories, were unable to find any evidence of blood on several "blood"-stained threads from the Shroud of Turin. The substance, which was suspiciously still red, failed sensitive tests for hemoglobin and hemoglobin derivatives, blood corpuscles, or any other identifiable blood components. The "blood" could not be identified as such, let alone be identified as to species or typed, and it contained reddish granules that would not even dissolve in reagents that dissolve blood. Sophisticated further tests—including microspectroscopic analysis, thin-layer chromatography, and neutron activation analysis—were also negative. Subsequently, famed microanalyst Walter McCrone identified the "blood" as tempera paint containing red ocher and vermilion along with traces of rose madder—pigments used by medieval artists to depict blood (Nickell 1998, 127–31).

Thus, when we are told that there is "human blood of the group AB" on the Oviedo "sudarium," and that this claim originates with Dr. Bollone, there is cause for skepticism. (Operating even further beyond his field of expertise, Bollone "has also studied the fabric of the sudarium, and affirmed that it is typical of the first century" [Guscin 1998, 56]—never mind seeking the opinion of textile experts.)

Another alleged correspondence between the "shroud" and the "sudarium" is that the "blood" stains on the latter supposedly "coincide exactly with the face of the image on the Turin Shroud." Dr. Alan Whanger claims to have found numerous "points of coincidence" between the Oviedo stains and the Turin image by employing a dubious overlay technique. Guscin (1998, 32) describes Whanger as a "highly respected scientist." Be that as it may, he is a retired geriatric psychiatrist and former missionary who has taken up image analysis as a hobby.

Whanger's judgment in such matters should perhaps be viewed in light of his studies of the Shroud of Turin. In photographs of that cloth's mottled image and off-image areas, Whanger has perceived, Rorschach-like, such crucifixion-associated items as "a large nail," "hammer," "sponge on a reed," "Roman thrusting spear," "loose coil of rope," pair of "sandals", "Roman dice," and numerous other imaginings. He and a botanist friend have also "identified" various "flower" images, as well as ancient Latin and Greek words such as "Jesus" and "Nazareth" (Nickell 2001).

Pollen "Evidence"

Still another purported link between the Turin and Oviedo cloths concerns pollen allegedly found on them. The shroud supposedly bears certain pollens characteristic of locales (Palestine, Constantinople, and ancient Edessa) that seemingly confirm a "theory" of the shroud's missing early history. Similarly, pollens supposedly discovered on the Oviedo Cloth seem to confirm its purported historical route (from Jerusalem through north Africa to Toledo and Oviedo); indeed, according to Guscin (1998, 22), they "perfectly match" the route. But perhaps the match is too good to be true.

The shroud pollens were reported by a Swiss criminologist, Max Frei-Sulzer, who, unfortunately, suffered severe credibility problems. Before his death in 1983, he represented himself as a handwriting expert and pronounced the notorious "Hitler diaries" authentic; they were proven forgeries soon thereafter. Frei's shroud pollen evidence was criticized on various grounds, especially when it was revealed that another shroud pollen sampling—taken at the same time as his—showed comparatively few pollen grains. As it turned out, so did Frei's, when his tape-lifted samples became available a few years after his death in 1983.

Walter McCrone was asked to authenticate the samples as coming from the Shroud of Turin. This he found easy to do, but he also discovered that the reputed pollens were essentially missing. An exception was one particular microscope slide that bore dozens of them in one small area—an indication, McCrone concluded, that the lift-off tape may have been peeled back from the slide and the pollen surreptitiously introduced. McCrone added (1993) that he had subsequently learned that Frei had been "censured by the Police hierarchy in Switzerland for . . . overenthusiastic interpretation of his evidence."

Unsurprisingly, the alleged pollen evidence that supposedly helps authenticate the Oviedo Cloth was also provided by Max Frei. In light of the suspicions raised about the shroud pollens, the Oviedo pollen claims should no longer be touted until an independent and impartial sampling is conducted.

Conclusions

As with the Shroud of Turin, the study of the Oviedo Cloth is obviously characterized by pseudoscience and possibly worse. The problems are symptomatic of the bias that can occur when analyses of a controversial object are conducted not by independent experts, chosen solely for their expertise, but instead by committed, self-selected partisans who begin with the desired answer and work backward to the evidence. As a result, science has once again been perverted in the interest of zealotry.

REFERENCES

Anderson, Mary Jo. 2000. Scientists: Relic authenticates Shroud of Turin, *World Net Daily*, 6 October. Last accessed 6 October 2000 from http://www.worldnetdaily.com

Guscin, Mark. 1998. *The Oviedo Cloth*. Cambridge, England: Lutterworth Press.

Humber, Thomas. 1978. *The Sacred Shroud*. New York: Pocket Books.

McCrone, Walter C. 1993. Letters to Joe Nickell, 11 June and 30 June.

———. 1996. *Judgement Day for the Turin "Shroud."* Chicago: Microscope Publications.

Nickell, Joe. 1998. *Inquest on the Shroud of Turin*. Amherst, N.Y.: Prometheus Books.

———. 2001. Scandals and follies of the "Holy Shroud." *Skeptical Inquirer* 25, no. 5: 17–20.
Whanger, Mary, and Alan Whanger. 1998. *The Shroud of Turin: An Adventure of Discovery*. Franklin, Tenn.: Providence House.
Wilson, Ian. 1979. *The Shroud of Turin*. Rev. ed. Garden City, N.Y.: Image Books.
———. 1998. *The Blood and the Shroud*. New York: Free Press.

28

A Typical Aries?

One of America's best-known astrologers was Linda Goodman. Revealingly, her own death raised serious questions about the validity of astrology.

Astrology is a means of fortunetelling. Those who believe in it claim it is a science by which a person's character, as well as his or her future, can be learned from the stars and planets. Depending on when one is born, a person supposedly comes under one of twelve astrological "signs." These signs are listed in Table 28-1, along with the main personality traits typically attributed to them.

In fact, however, astrologers often assign contradictory traits to a sign. For example, we are told of Gemini people that they are "twins," but that "the two sides of their character war with each other." Taurians (Taurus people) are purportedly "agreeable" in their home life, yet we

TABLE 28-1. **The 12 signs of the zodiac together with the traits typically ascribed to each.**

Date of birth	*Sign (and symbol)*	*Traits*
1. Mar. 22–Apr. 20	Aries (Ram)	ambitious, idealistic, high-spirited, generous
2. Apr. 21–May 21	Taurus (Bull)	home-loving, usually calm, fond of creature comforts
3. May 22–June 21	Gemini (Twins)	brilliant, changeable, charming
4. June 22–July 23	Cancer (Crab)	sensitive, nervous, home-loving
5. July 24–Aug. 23	Leo (Lion)	big, strong, regal (kingly)
6. Aug. 24–Sept. 23	Virgo (Virgin)	more intellectual than emotional, self-critical
7. Sept. 24–Oct. 23	Libra (Scales)	sociable, talented, frequently artistic
8. Oct. 24–Nov. 22	Scorpio (Scorpion)	passionate, whether temperamental or hard-working
9. Nov. 23–Dec. 22	Sagittarius (Archer)	frank, fearless, loyal, unselfish
10. Dec 23–Jan. 20	Capricorn (Goat)	tenacious, able to overcome obstacles
11. Jan. 21–Feb. 19	Aquarius (Waterbearer)	interested in others, giving, loyal
12. Feb. 20–Mar. 21	Pisces (Fishes)	generous, popular, tend to be dreamers

are warned: "It is never safe to wave the red flag in front of the bull!" In the same vein, Libras are called easygoing, yet "they have it in them to take a firm stand."

Conversely, similar characteristics are also assigned to different signs. For instance, both Cancers and Taurians are said to be fond of the home. Sagittarians are "loyal," just like Aquarians, and both Sagittarians and Aries people are "idealists" (Adams 1931).

Actually, since the system of astrology was created in ancient times, the earth's position in relation to the planets has shifted. Most astrologers ignore this fact. Also, astrologers treat the "influences" of the planets as equal, although the planets are at vastly different distances from earth. Putting aside these objections, one can still ask, as astrology critic Milbourne Christopher (1970, 101–14) did:

> Why is someone born at a certain time, on a certain day, in a certain part of the world of a certain nature? Did the originators of astrology study the traits of millions of people and discover that all those born at a specific time had identical characteristics and futures? No! Nor has it been proven since that the early fictions are fact.

Linda Goodman was an arch-promoter of astrology. An itinerant newspaper and radio writer, she adopted her pseudonymous first name after a stint on radio reading "Letters from Linda." Her surname was that of her second husband, Sam Goodman.

Goodman's interest in astrology came from supermarket booklets, and her own books took advantage of the extreme popularity of the subject in America. In 1968, Goodman's *Sun Signs* became the first book on the topic to make the *New York Times* best-seller list, and she received $2.3 million for the paperback rights alone. In time, she became even more mystical and incorporated both numerology and reincarnation concepts into her books.

When her 18-year-old daughter Sarah committed suicide in 1973, Goodman's reaction was to refuse to believe that the body her husband had identified was actually Sarah's, citing as "evidence" her daughter's horoscope. Instead, she embarked on a search for the "missing" teenager, squandering her money, and (according to her obituary in the October 25, 1995, *New York Times*), "living for several months on the steps of St. Patrick's Cathedral."

According to the *Times*, Goodman was born Mary Alice Kemery [*sic*] in Morgantown, West Virginia, "on a date she gave as April 19 in a year she would never disclose"—not even to her son. This secrecy prompted a minor investigation: I commissioned a professional genealogical record searcher to determine the facts relating to her birth. Actually, according to the Monongalia County, West Virginia, Register and Index of Births, she was Mary Alice *Kenery*, a daughter of Robert S. and Mazy A. (McBee) Kenery. She was born in 1925 (the *Times* guessed she was "about 70") but not on the day she alleged. The register lists her birth as April 10, nine days earlier than Goodman always claimed. Therefore, from an astrological point of view, her horoscope would have been significantly different than she represented it (although it is not certain that this was a deliberate falsification on her part) (Nickell 1996).

In any case, her forecast for the day of her death (as given in Jeane Dixon's syndicated *Horoscope* column) read, in part: "A change of scenery will help you put recent events in proper perspective."

REFERENCES

Adams, Evangeline. 1931. *Astrology for Everyone.* New York: New Home Library.

Christopher, Milbourne. 1970. *ESP, Seers & Psychics.* New York: Thomas Y. Crowell.

Nickell, Joe. 1996. A typical Aries? *Skeptical Inquirer* 20, no. 1 (January/February): 5.

29

The Case of the Psychic Shamus

Do Psychics Really Help Solve Crimes?

Among the cases of alleged psychic sleuthing that I have investigated, some have seemed impressive at first sight, but all eventually succumbed to the light of inquiry. One example stemmed from my appearance on the *Mark Walberg Show*, televised 7 February 1996, with a self-proclaimed psychic named Ron Bard. He boasted that he had "solved over 110 murder cases"; when asked to name one, he cited the case of two girls in Harrison, New York. Bard claimed that a key on one girl's body led him to the South Bronx where "the key worked in the lock and that's how we found the murderer." Actually, however, as the Harrison police chief later told me, Bard was not consulted by the police; the only key in the case was one that turned out to belong to one of the victims; and the case was solved by "diligent police work, not visions" (Nickell

2001, 210–13). (For another example of a case that collapsed under scrutiny, see chapter 9, "Remotely Viewed? The Charlie Jordan Case.")

Hold on, though: What about repeated testimonials from experienced homicide detectives regarding a specific psychic—like Nutley, New Jersey's Dorothy Allison? Was she indeed a psychic shamus, or just a psychic sham?

I included Allison in a book I edited called *Psychic Sleuths*, featuring her in an in-depth critical study by investigative writer Michael R. Dennett (1994). I later debunked her claims in such media venues as *Dateline NBC*, various radio shows, Internet postings, other articles and books, and newspaper interviews. For good measure, I was even quoted in her obituary in the *New York Times*, prompting one colleague to joke that I had hounded her to the grave.

Allison died December 1, 1999 (my fifty-fifth birthday). In her obituary, the *Times* observed that Allison was unsuccessful in solving the Patty Hearst kidnapping, the "Son of Sam" killings, the Atlanta child murders, and the murder of JonBenet Ramsey—all cases into which Allison insinuated herself (Martin 1999).

JonBenet Ramsey was the child beauty princess whose tiny body was found bludgeoned and strangled in her parents' Boulder, Colorado, home on Christmas night, 1996. In this instance, Allison made bold pronouncements on the 27 April1998, airing of *LEEZA* (the Leeza Gibbons show).

Such grandstanding is not surprising. Psychics thrive on the media attention they can get from high-profile cases. I was tipped to the show by a letter, written the following day, from my friend, the great entertainer and author, Steve Allen. "Dear Joe," he began. "A quick note, while on the usual run." Until his death in 2000, Steve was a dynamo, with numerous activities that included serving as co-chair of the Council for Media Integrity, which was a branch of my employer, the Committee for Scientific Investigation of Claims of the Paranormal (CSICOP). Although we corresponded about various things, Steve was obviously wearing his Council hat on this occasion, as he suggested that I obtain a transcript of the *LEEZA* show for study.

In his note he had referred to Allison only as "an alleged psychic who it was said is consulted by some police departments in connection with unsolved murders, such as the JonBenet case." I dutifully obtained a video copy of the program, and I was not surprised that the psychic was Allison. On camera, in her typically overstated manner, she insisted that the little girl's parents were "absolutely not" involved, and that the real killer was a former handyman. She perceived "connections" to Germany and Georgia, the numbers 2–8–9, and the names "Martin" and "Irving"—the latter, she said, being "the one I think that did this" (the murder). Working with a police artist, Allison produced a drawing of the alleged killer.

Leeza emphasized that Allison had gone far out on the proverbial limb, and some audience members seemed quite skeptical of the clairvoyant's "clues." One woman challenged Allison to tell *where* the alleged murderer was, as Leeza tactfully took the opportunity to go to a commercial break.

None of this seemed to be of any use to law enforcement. I asked Steve Allen to make a brief statement that I could include in an Internet posting. He wrote:

> The important question, in cases of this sort, is to determine whether a) the alleged psychic is a bare-faced liar or b) honestly self-deluded, in the way that many religious fanatics are. In the meantime television producers must be urged to consult scientists and other authorities who are perfectly aware of the essential absurdity of claims made by psychics, fortunetellers, tarot-card readers, or astrologers. It is relevant to quote here a little quatrain I wrote years ago in an album for children titled "How to Think."

Look for the evidence.
Look for the proof.
Or else you're acting
like an awful goof.

In my subsequent article, I pointed out that although Allison was billed as a "police psychic who helped solve 5,000 cases," the truth is quite different. What self-styled psychics like Allison typically do is simply toss out some vague statements that they call "clues." Those invariably prove meaningless until after the crime is solved or the victim is located, whereupon the psychic interprets the statements in light of the facts. As an unimpressed Georgia police chief summed up a case on which Allison had made pronouncements: "She said a whole lot of things, a whole lot of opinions, partial information and descriptions. She said a lot. If you say enough, there's got to be something that fits." Most of her reported successes appeared to be like the one that a New Jersey police captain attributed to her. Her predictions "were difficult to verify when initially given," he said. "The accuracy usually could not be verified until the investigation had come to a conclusion" (quoted in Dennett 1994, 46). Although Allison convinced many reporters and even police that she had a criminological sixth sense, in fact she used a tried-and-true formula: Arrive on the scene of a high-profile case, make numerous vague pronouncements, and then, after the true facts become known, interpret the statements accordingly—a technique known as *retrofitting*.

For example, a 1997 episode of TV's *Crackdown on Crime* said of Allison, "Nutley police asked her to find a missing five-year-old boy. She did. He had drowned in a pipe during a storm." In fact, however, every one of those statements is untrue. It was Allison who approached the police to share a "vision." Not only did she fail to locate the child's body, but she also caused police to waste considerable time and resources following up on her hints. The boy's body was actually discovered later, floating in a pond, by a man seeking a spot to bury his dead cat. But through the technique of retrofitting, Allison converted her failure into a seeming success, mentioning details of the boy's clothing that she had supposedly "seen" accurately.

For another instance, consider the case of murder victim Susan Jacobson. Jacobson's body, discovered in 1976 on Staten Island, was found neither by Allison nor by the police, but rather by a 13-year-old boy who had been playing with friends. Allison's prior statement that she had clairvoyantly seen "horses along a trail" was subsequently interpreted as a "hit" because the cemetery where the victim was finally laid to rest had, Allison stated, "once been a bridle path" (Dennett 1994).

Psychics have other means of scoring apparent successes, including making exaggerated or false claims about previous cases to uncritical reporters, shrewdly studying local newspaper files and area maps, gleaning information from family members or others associated with a tragedy, and even impersonating police and reportedly attempting to bribe detectives. Some credulous police officers even help the psychic in the reinterpretation necessary to convert a failure into an apparent success. (For example, in one case where there was no nearby church, as had been predicted, property owned by a church was counted as fitting the criterion.) The result is like painting a bull's-eye around an arrow after the arrow has lodged somewhere (Nickell 1994).

It is not just that "psychics" are of no help to law enforcement; they actually cause wasted resources, as indicated earlier in the case of the missing five-year-old boy. The Nutley, New Jersey, police spent a whole afternoon digging up a drainage ditch that Dorothy Allison mistakenly thought contained the body of the missing child.

Psychic shamus or sham? The evidence speaks for itself.

REFERENCES

Dennett, Michael R. 1994. America's most famous psychic sleuth: Dorothy Allison. In *Psychic Sleuths: ESP and Sensational Cases*, edited by Joe Nickell, 42–59. Buffalo, N.Y.: Prometheus Books.

Martin, Douglas. 1999. Dorothy Alison, 74, "psychic detective" consulted by police. *New York Times*, 20 December.

Nickell, Joe, ed. 1994. *Psychic Sleuths: ESP and Sensational Cases*. Buffalo, N.Y.: Prometheus Books.

———. 2001. *Real-life X-files: Investigating the Paranormal*. Lexington, Ky.: University Press of Kentucky.

30

The Pagan Stone

A carryover from the past that is finding new expression in today's Russia is devotion to a certain ancient stone that was a focus of pagan ceremonies and is reputed to have magical properties ("Kolomenskoye" n.d.). In October 2001, I was able to visit the site with the assistance of a young woman named Nadya Tereshina. (A reception supervisor at my hotel, she was willing to be helpful while at the same time taking the opportunity to practice her English.)

The stone is in Kolomenskoye, a state museum-reserve in south Moscow. Located on the west bank of the Moskva River, Kolomenskoye was once a royal summer retreat of Czar Ivan the Terrible (1530–1584). A guidebook calls it "one of the most evocative sites in Moscow" (Richardson 1998, 239), for its variety of interesting aspects. It contains an eleventh-century Slavic archaeological site; woods of hoary oaks (one 600 years old); the Church of the Ascension (built 1532), which houses the Great Thaumaturgic Russian Icon of "The Virgin Majestic"; and, among many other treasures, the rustic log cottage of Czar Peter the Great (1672–1725).

Muscovites flock to Kolomenskoye for festivals, summer sunbathing, and other activities. At one mid-January festival, children dress up as horses, bears, or ghosts to sing carols and give blessings in exchange for gifts. Another festival is based on a pagan spring celebration that later marked the beginning of Lent. Still other celebrations have folkloric, cultural, or religious themes ("Kolomenskoye" n.d.; Richardson 1998, 239–46).

The purportedly magical stone is actually one of two huge sandstone boulders left by continental glaciation and found a short distance apart in a spring-studded ravine of Kolomenskoye. Reportedly once serving "for pagan tribes' ritual ceremonies," they are dubbed the "Goose" and "Girlish Stone" ("Kolomenskoye" n.d.), with the latter reposing on a hillside. As luck would have it, when we arrived on a Sunday afternoon, following a sketchy map provided by one of Nadya's co-workers, Kirill Boliakiu, there was a ceremony in progress at that stone (Figure 30-1).

The ceremony was not pagan—there were references to the "Christian family"—but it seemed to me to be . . . well, *paganesque* (to coin a term). The leader of the group, wearing a belted robe and sporting long hair and a beard, had a decidedly New-Age appearance. He cleaned the stone and partially covered it with an animal pelt that had been draped over his shoulder, to "warm" the rock. Crusts of bread were scattered on the stone, because it was supposed that when birds carried the offerings away, the scatterer's wishes would be granted. The supplicants raised their hands skyward, in the ancient attitude of prayer, as the leader gave an invocation. Finally, he strummed his zither and sang a song about "Old Russia."

Poignant as such a ceremony may seem, there is no evidence that the stone has any magical power or functions as anything other than a focus for people's hopes. Petitions, including those for cures of illness, appear to be granted (or not) according to the whim of circumstance

FIGURE 30-1. **New-Age religious ceremony at a boulder once used for rituals by pagan tribes.**

and possibly the beseecher's own subsequent actions. The wishes that are fulfilled are accepted as proof of the stone's power, whereas those that are not are forgotten or explained away.

Suppressed during the communist era, religion is again flourishing in Russia as part of the new climate of change and tolerance of individual expression. Cults and offbeat groups like the one worshipping at the ancient stone—with its admixture of pagan, vaguely Orthodox, and New-Age elements—are one manifestation of the trend.

REFERENCES

"Kolomenskoye: The State Museum Reserve." N.d. [current 2001]. Brochure.
Kruglyakov, Edward. 2001. Personal communication, 6 April and 3–5 October.
Richardson, Dan. 1998. *Moscow: The Rough Guide*. London: Rough Guides, Ltd.

31

Benny Hinn

Healer or Hypnotist?

Benny Hinn tours the world with his "Miracle Crusade," drawing thousands to each service, with many hoping for a healing of body, mind, or spirit. A significant number seem to be rewarded and are brought onstage to pour out tearful testimonials. Then, seemingly by the Holy Spirit, they are knocked down at a mere touch or gesture from the charismatic evangelist. Although I had seen clips of Hinn's services on television, I decided to attend and witness his performance live when his crusade came to Buffalo, New York, for June 28–29 of 2001. Donning suitable garb and sporting a cane (left over from a 1997 accident in Spain), I limped into my seat at the HSBC Arena in downtown Buffalo.

Learning the Ropes

Benny Hinn was born in 1953, the son of an Armenian mother and a Greek father. He grew up in Jaffa, Israel, "in a Greek Orthodox home," but was "taught by nuns at a Catholic school" (Hinn 1999, 8). Following the Six-Day War in 1967, he emigrated to Canada with his family. When he was 19, he became a born-again Christian. Nearly two years later, in December 1973, he traveled by charter bus from Toronto to Pittsburgh to attend a "miracle service" by Pentecostal faith-healing evangelist Kathryn Kuhlman (1907–1976). At that service he had a profound religious experience, and that very night he was pulled from bed and "began to shake and vibrate all over" with the Holy Spirit (Hinn 1999, 8–14).

Before long, Hinn began to conduct services sponsored by the Kathryn Kuhlman Foundation. Kuhlman died before Hinn could meet her personally, but her influence on him was extensive, as he acknowledged in a book, *Kathryn Kuhlman: Her Spiritual Legacy and Its Impact on My Life* (Hinn 1999). Eventually he began preaching elsewhere, including the Full Gospel Tabernacle in Orchard Park, New York (near Buffalo) and later at a church in Orlando, Florida. By 1990, he was receiving national notice for his book *Good Morning, Holy Spirit*, and in 1999 he moved his ministry headquarters to Dallas, Texas.

Lacking any biblical or other theological training, Hinn was soon criticized by other Christian ministries. One, Personal Freedom Outreach, labeled his teachings a "theological quagmire emanating from biblical misinterpretation and extra-biblical 'revelation knowledge.'" He admitted to *Christianity Today* magazine that he had erred theologically and vowed to make changes (Frame 1991), but he remains controversial. Nevertheless, according to a minister friend, "[o]utside of the Billy Graham crusade, he probably draws the largest crowd of any evangelist in America today" (Condren 2001).

Hinn's mentor, Kathryn Kuhlman, who performed in flowing white garments trimmed with gold (Spraggett 1971, 16), was apparently the inspiration for Hinn's trademark white suits and gold jewelry. From her he obviously learned the clever "shotgun" technique of faith-healing (also practiced by Pat Robertson and others). This involves announcing to an audience that certain healings are taking place, without specifying just who is being favored (Randi 1987, 228–229).

Selection Process

In employing this technique, Hinn first sets the stage with mood music, leading the audience (as did Kuhlman) in a gentle rendering of

> He touched me, oh, He touched me,
> And, oh, the joy that filled my soul!
> Something happened and now I know
> He touched me, and made me whole. . . .

Spraggett (1971, 17) says that with Kuhlman, as it was sung over and over, the song became "a chant, an incantation, hypnotic in its effect"; the same is true of Hinn's approach.

In time, the evangelist announces that miracles are taking place. At the service I attended, he declared that someone was being "healed of witchcraft"; others were having the "demon of suicide" driven out; still others were being cured of cancer. He named various diseases and conditions that were supposedly being alleviated and mentioned different areas of the anatomy—a back, a leg, etc.—that he claimed were being healed. He even stated that he did not need to name every disease or body part, that God's power was effecting a multitude of cures all over the arena.

Thus, instead of the afflicted being invited up *to be* healed (with no guarantee of success), the "shotgun" method encourages receptive, emotional individuals to believe that they *have been* healed. Only that self-selected group is invited to come forward and testify to their supposedly miraculous transformations. Although I remained seated (seeing no investigative purpose in making a false testimonial), others were more tragically left behind. At one Hinn service, a woman, hearing the evangelist's anonymously directed command to "stand up out of that wheelchair!," struggled to do so for almost half an hour before finally sinking back into it, exhausted (Thomas 2001).

There is even a further step in the selection process: Of those who do make it down the aisles, only a very few will actually be invited onto the stage. They must first undergo what amounts to an audition for the privilege. Those who tell the most interesting stories and show the greatest enthusiasm are the ones most likely to be chosen (Underdown 2001).

Perhaps not surprisingly, this selection process is virtually identical to that employed by professional stage hypnotists. According to Robert A. Baker, in his definitive book, *They Call It Hypnosis* (1990, 138–39):

> Stage hypnotists, like successful trial lawyers, have long known their most important task is to carefully pick their subjects—for the stage as for a jury—if they expect to win. Compliance is highly desirable, and to determine this ahead of time, the stage magician will usually give several test suggestions to those who volunteer to come up on the stage. Typically, he may ask the volunteers to clasp their hands together tightly and then suggest that the hands are stuck together so that they can't pull them apart. The stage hypnotist selects the candidates who go along with the suggestion and cannot get their hands apart until he tells them, "Now, it's okay to relax and separate them." If he has too many candidates from the first test, he may then give them a second test by suggesting they cannot open their mouths, move a limb, or open their eyes after closing them. Those volunteers who fail one or more of the tests are sent back to their seats, and those who pass all the tests are kept for the demonstration. Needless to say, not only are they compliant, cooperative, and suggestible, but most have

already made up their minds in volunteering to help out
and do exactly as they are told.

Role-Playing

Once on stage, one of Hinn's screeners announces each "healed" person in turn, giving a quick summary of the alleged miracle. At the service I attended, one woman put on a show of jumping up and down to demonstrate that she was free of pain following knee surgery three weeks before. Another was cured of "depression"—caused by "The Demon," said a screener—that resulted from "an abusive relationship with her husband." Still another (who admitted to being "an emotional person") said that her sister-in-law sitting beside her had begun to "speak in tongues," and that she herself felt she had been healed of various ailments, including high blood pressure and marital trouble. At the mention of her brother, Hinn brought him up, whereupon the healer learned that the brother had been healed of "sixteen demons" two years previously, and expected to be cured of diabetes. Hinn prayed for God to "set him free" of the disease. Another was supposedly cured of being "afraid of the Lord" (although he was carrying the Bible of a friend who had died of AIDS), and one woman stated that she believed she had just been cured of ovarian cancer.

In each instance—after the person has given a little performance (running about, offering a sobbing testimonial, or the like) and Hinn has responded with some mini-sermon, prayer, or other reaction—the next step in the role-playing is acted out. As one of his official catchers moves into place behind the person, Hinn gives a gesture, touch, or other signal. Most often, while squeezing the person's face between thumb and finger, he gives a little push, and down the compliant individual goes. Some slump; some stiffen and fall backward; a few reel. Once down, many lie as if entranced, whereas others writhe and seem almost possessed.

Along with speaking or praying in tongues (*glossolalia*) and other emotional expressions, this phenomenon of "going under the Power" is a characteristic of the modern charismatic movement (after the Greek *charisma*, "gift"). Also known as being "slain in the Spirit," it is often regarded skeptically even by other Christians, who suspect—correctly—that the individuals involved are merely "predisposed to fall" (Benny Hinn: Pros & Cons 2002). That is, they merely engage in a form of role-playing that is prompted by their strong desire to receive divine power and by the influence of suggestion that they do so. Even the less emotionally suggestible people will be unwilling *not* to comply when those around them expect a certain behavior.

In short, they act just as if they had been "hypnotized." Although popularly believed to involve a mystical "trance" state, hypnosis is in fact just compliant behavior in response to suggestions (Baker 1990, 286). One professional hypnotist said of Hinn's performance: "This is something we do every day and Mr. Hinn is a real professional" (Thomas 2001).

Cures?

What about the healings? Do faith-healers like Benny Hinn really help nudge God into working miracle cures? In fact, such claims are invariably based on negative evidence—"we don't know what caused the illness to abate, so it must have been supernatural"—and thus are examples of the logical fallacy called "arguing from ignorance." In fact, as I explained to a reporter from *The Buffalo News* following a Benny Hinn service, people's feelings that they have been healed are due to several factors. In addition to the body's own natural healing mechanisms, there is the fact that some serious ailments, including certain types of cancer, are unpredictable and may undergo "spontaneous remission"; that is, they may abate for a time or go away entirely. Other factors include such egregious occurrences as misdiagnosis, such as that of a supposedly "inoperable, malignant brain-stem tumor" that was actually an artifact of a faulty CT scan (Randi 1987, 291–92).

And then there are the powerful effects of suggestion. Not only psychosomatic illnesses (of which there is an impressive variety), but also those with distinct physical causes may respond to some degree to "mental medicine." Pain is especially responsive to suggestion. In the excitement of an evangelical revival, the reduction of pain due to the release of endorphins (pain-killing substances produced by the body) often causes people to believe that, and act as if, they have been miraculously healed (Condren 2001; Nickell 1993; Nolen 1974).

Critical studies are illuminating. Dr. William A. Nolen, in his book *Healing: A Doctor in Search of a Miracle* (1974), followed up on several reported cases of healing from a Kathryn Kuhlman service but found no miracles—only remissions, psychosomatic diseases, and other natural explanations, including the power of suggestion.

More recently, a study was conducted following a Benny Hinn crusade in Portland, Oregon, where 76 miracles were alleged. For an HBO television special, *A Question of Miracles* (Thomas 2001), Benny Hinn Ministries was asked to supply the names of as many of these as possible for investigation. After 13 weeks, just 5 names were provided. Each case was followed for one year.

The first was that of a grandmother who stated that she had had "seven broken vertebras," but that the Lord had healed her at the evening service in Portland. In fact, subsequent x-rays revealed otherwise, although the woman felt that her pain had lessened.

The second case involved a man who had suffered a logging accident 10 years earlier. He demonstrated improved mobility at the crusade, but his condition afterward deteriorated and "movement became so painful he could no longer dress himself." Nevertheless, he remained convinced that he had been healed, and thus refused the medication and surgery his doctors insisted were necessary.

The next individual was a lady who, for 50 years, had only "thirty percent of her hearing," as claimed at the Portland crusade. However, her physician stated, "I do not think this was a miracle in any sense." He reported that the woman had had only a "very mild hearing loss" just two years before, and that she had made "a normal recovery."

The fourth case was that of a girl who had not been "getting enough oxygen" but who claimed to have been healed at Hinn's service. In fact, since the crusade she had "continued to suffer breathlessness"—yet her mother was so convinced that a miracle had occurred that she did not continue to have her daughter seek medical care.

Finally, there was what the crusade billed as "a walking dead woman." She had had cancer throughout both lungs, but her doctors were now "overwhelmed" that she was "still alive and still breathing." Actually, her oncologist rejected all such claims, declaring that the woman had an "unpredictable form of cancer that was stable at the time of the crusade." Tragically, her condition subsequently deteriorated and she died just nine months afterward.

What Harm?

As these cases demonstrate, there is a danger that people who believe themselves cured will forsake medical assistance that could bring them relief or even save their lives. Dr. Nolen (1974, 97–99) relates the tragic case of Mrs. Helen Sullivan, who suffered from cancer that had spread to her vertebrae. Kathryn Kuhlman had her get out of her wheelchair, remove her back brace, and run across the stage repeatedly. The crowd applauded what they thought was a miracle, but the antics cost Mrs. Sullivan a collapsed vertebra. Four months after her "cure," she died.

Nolen (1974, 101) stated that he did not think Miss Kuhlman a deliberate charlatan. She was, he said, ignorant of diseases and the effects of suggestion, but he suspected she had "trained herself to deny, emotionally and intellectually, anything that might threaten the validity of her ministry." The same may apply to Benny Hinn. One expert in mental states, Michael A. Persinger, a neuroscientist, suggests that people like Hinn have fantasy-prone personalities (Thomas 2001). Indeed, the backgrounds of both Kuhlman and Hinn reveal many traits associated with fantasy-proneness, but it must be noted that being fantasy-prone does not preclude also being deceptive and manipulative.

Hinn notes that only rarely does he lay hands on someone for healing. He made an exception, however, for one child whose case was being filmed for the HBO documentary. The boy was blind and dying from a brain tumor. "The Lord's going to touch you," Hinn promised. The child's parents believed and, although they were not wealthy, pledged $100 per month to the Benny Hinn Ministries. Subsequently, however, the child died.

Critics like the Rev. Joseph C. Hough, president of New York's Union Theological Seminary, say of the desperately hopeful: "It breaks your heart to know that they are being deceived, because they genuinely are hoping and believing. And they'll leave there thinking that if they didn't get a miracle it's because they didn't believe." More pointedly, Rabbi Harold S. Kushner stated, during an interview on *A Question of Miracles* (Thomas 2001):

> I hope there is a special place in Hell for people who try and enrich themselves on the suffering of others. To tantalize the blind, the lame, the dying, the afflicted, the terminally ill, to dangle hope before parents of a severely afflicted child, is an indescribably cruel thing to do, and to do it in the name of God, to do it in the name of religion, I think, is unforgivable.

Amen.

REFERENCES

Baker, Robert A. 1990. *They Call It Hypnosis*. Buffalo, N.Y.: Prometheus Books.

Benny Hinn: Pros & cons. 2002. Internet posting. Retrieved 11 January from www.rapidnet.com/~jbeard/bdm/exposes/hinn/general.htm

Condren, Dave. 2001. Evangelist Benny Hinn packs arena. *Buffalo News* (Buffalo, N.Y.), 29 June.

Frame, Randy. 1991. Best-selling author admits mistakes, vows changes. *Christianity Today*, 28 October, 44–45.

Hinn, Benny. 1990. *Good Morning, Holy Spirit.* Nashville, Tenn.: Thomas Nelson.

———. 1999. *Kathryn Kuhlman: Her Spiritual Legacy and Its Impact on My Life.* Nashville, Tenn.: Thomas Nelson.

Nickell, Joe. 1993. *Looking for a Miracle: Weeping Icons, Relics, Stigmata, Visions & Healing Cures.* Amherst, N.Y.: Prometheus Books.

Nolen, William A. 1974. *Healing: A Doctor in Search of a Miracle.* New York: Random House.

Randi, James. 1987. *The Faith Healers.* Buffalo, N.Y.: Prometheus Books.

Spraggett, Allen. 1971. *Kathryn Kuhlman: The Woman Who Believed in Miracles.* New York: Signet.

Thomas, Antony. 2001. *A Question of Miracles.* HBO special, aired on 15 April.

Underdown, James. 2001. Personal communication with author, 23 October.

32

Australia's Convict Ghosts

It is a spectacular land to which many superlatives apply. Although not, as often claimed, the oldest of the continents (the cores of which are approximately the same age), Australia *is* the world's smallest continent and—excepting Antarctica—the flattest and driest one. Separated from the other continents for some 40 million years, Australia has produced unique flora and fauna, and its history "began twice": first, some 50,000 to 60,000 years ago when the nomadic Aborigines reached the shores; and second, on January 18–20, 1788, when 11 British ships arrived laden with convicts (Chambers 1999, 1–10).

I had a wonderful opportunity to visit Down Under during the Third Skeptics World Convention, held in Sydney on 10–12 November 2000. I determined to extend my sojourn another two weeks, so that I could investigate several myths and mysteries. I began with the "haunted" Hyde Park Barracks (Figure 32-1).

FIGURE 32-1. **Old Hyde Park Barracks was constructed to house male convicts. (Photograph by Joe Nickell.)**

Reputedly "the most haunted building in Central Sydney" (Davis 1998, 2), the Hyde Park Barracks was constructed in 1817 as secure housing for government-assisted male convicts. Opened in mid-1819, its central building held an average of 600 men, who were assigned to various workplaces by day and lodged at night in 12 rooms outfitted with hammocks (Figure 32-2). (In 1848 it was transformed into an immigration depot for single females, and in 1887 it became a government office complex. It is now a museum featuring its original history.)

FIGURE 32-2. **Hammocks in the "haunted" barracks in Sydney where it is possible to spend the night, thus recreating the "convict experience." (Photograph by Joe Nickell.)**

Ghosts went unreported at the Barracks until the 1950s, when a clerk claimed to have seen the apparition of a "figure in convict garb hobbling down a corridor" (Davis 1998, 2). Since the building became a museum, it has been the focus of many reports of various phenomena, attested to by security guards and others who spend the night there, including schoolchildren who stay on organized sleepovers to gain the "convict experience."

Unlike most "haunted" places, where ghostly shenanigans are typically available only as hand-me-down stories, "it is said that" tales, or anecdotes collected for the obligatory Halloween newspaper article or similar entertainment, the Barracks maintains a ghost file, containing accounts of experiences recorded just after they occurred. Curator Michael Bogle graciously made these available for me to study in his office.

Bogle takes a professionally neutral stance on the subject of hauntings, but admits that he has himself has had no ghostly experiences. Neither had four other staff members I interviewed there; a fifth described a few incidents that she attributed to a ghost, but none of those occurred at the Barracks.

Despite the neutrality, the museum's solicitation of overnight visitors' "thoughts and feelings" about their visit—utilizing a handout with space to record their impressions—no doubt encourages spooky thoughts. The handout says in part: "Should you have an 'eerie' meeting of some sort, or merely sense an inexplicable presence, the museum would appreciate your description—with as much detail as possible." It continues: "The accompanying [floor] plans will help you on your journey through the building and enable you, where appropriate, to map any 'out of the ordinary' occurrences." The place where the overnighters slumber on the third floor is called the "sleep and dream" area (again see Figure 32-1). Obviously, the entire experience is designed to stimulate the imagination, perhaps provoking dreams or even triggering apparitions of figures from the past.

Not surprisingly, then, several people did report having eerie feelings, in which suggestibility no doubt played a role. For instance, one pre-Halloween (11 October) account from 1991 stated that a security guard "hoped" a certain fellow guard "could make a connection with the ghost," which "everyone in Security knew of" and which was typically experienced as "a chilling sensation" on the third floor.

Other respondents described apparent "waking dreams," the sometimes apparitional experiences that occur in the twilight between wakefulness and sleep and that may also involve *sleep paralysis*, an inability to move because the body is still in the sleep mode (Nickell 1995; 2000). For example, one respondent reported seeing "a man standing beside my hammock looking at me" and wearing period clothes. Her account reveals she had "tried to imagine what it must have been like for the convicts who stayed there," thus helping set the stage for such an experience.

Another woman, upon going to sleep, felt "a massive pressure upon my body and a stark feeling that something was trying to take over"—an experience entirely consistent with sleep paralysis.

On occasion, the written narratives contain suggestions of possible pranking, as with the one of 47 schoolchildren who felt a "long hand" reach in under her sleeping bag to touch her on the hip (or was that instead merely the effect of a runaway imagination, or even another waking dream?). Once, a child's footsteps, heard by two guards, were first attributed to one of the children having gotten up; as that reportedly turned out not to have been the case, the incident was explained as a sound that "must have been made by the wind." One experiencer heard a tapping sound that staff subsequently ascribed to a mechanized display.

Such incidents seem typical of those reported at the Hyde Park Barracks, as well as at many other allegedly haunted sites. For instance, "some say" that the Old Melbourne Gaol is "the repository of many troubled spirits, the ghosts of criminals who suffered and died there" (Davis 1998, 174). Certainly it is a stark showing of nineteenth-century penal life, with exhibits of grim implements of restraint and punishment and various *mementi mori*. Examples include the death mask, pistol, and homemade armor of the notorious "bushranger" (highwayman) Ned Kelley (the armor was effective until Kelly was shot in the knee). It also includes the scaffold on which Kelly was subsequently hanged, following his ironic final words, "Such is life." In short, the gaol is one of those places that, if not actually haunted, certainly *ought* to be. An advertising brochure promises: "Experience the haunting and eerie atmosphere of the gaol, and by listening carefully, you can almost hear the clank of the prisoners' chains."

However, hard evidence of ghostly phenomena at the site is scant, notwithstanding a questionable "ghost" photo halfheartedly brought out by a gift-shop employee when the topic of hauntings was broached. She conceded that some people did get "feelings" at the site, but noted that she had worked there for 10 years without any paranormal experiences of her own. She jokingly pointed out that she only worked one day a week and that perhaps "the ghosts take Tuesdays off."

REFERENCES

Chambers, John H. 1999. *A Traveler's History of Australia.* Gloucestershire, U.K.: Windrush Press.

Davis, Richard. 1998. *The Ghost Guide to Australia.* Sydney: Bantam Books.

Nickell, Joe. 1995. *Entities: Angels, Spirits, Demons, and Other Alien Beings.* Amherst, N.Y.: Prometheus Books.

———. 2000. Haunted inns: Tales of spectral guests. *Skeptical Inquirer* 24, no. 5 (September/October): 17–21.

33

Psychic Pets and Pet Psychics

Many believe that the bond between man and animals, known from great antiquity, includes extrasensory perception (ESP). They cite anecdotal evidence, controversial research data, and the claims of alleged psychics. During more than three decades of investigating the paranormal, I have often encountered and reviewed such evidence. I have written about "talking" animals, appeared with a "pet psychic" on *The Jerry Springer Show*, analyzed alleged paranormal communications between people and animals (both living and dead), and even visited a spiritualists' pet cemetery. Here is a look at some of what I have found.

"Talking" Animals

Alleged animal prodigies—various "educated," "talking," and "psychic" creatures—have long been exhibited. In seventeenth-century France, for instance, a famous "talking" horse named Morocco seemed to possess such remarkable powers, including the ability to do mathematical calculations, that he was charged with "consorting with the Devil." However, he saved his own and his master's life when he knelt, seemingly repentant, before church authorities.

In the latter eighteenth century, a "Learned Pig" and a "Wonderful Intelligent Goose" appeared in London. The porker spelled names, solved arithmetic problems, and even read thoughts by selecting, from flashcards, words thought of by audience members (Jay 1986, 7–27). The goose, advertised as "The greatest Curiosity ever witnessed," performed such feats as divining a selected playing card, discovering secretly selected numbers, and telling time "to a Minute" by a spectator's own watch (Christopher 1962, 35).

Other prodigies were Munito the celebrated dog, Toby "the Sapient Pig" (who could "Discover a Person's Thoughts"), and a "scientific" Spanish pony who shared billing with "Two Curious Birds." The latter were "much superior in knowledge to the Learned Pig" and "the first of the kind ever seen in the World." Such animals typically performed their feats by stamping a hoof or paw a certain number of times or by spelling out answers using alphabet and number cards (Christopher 1962, 8–37; Jay 1986, 7–27).

In 1904, a German horse named Clever Hans provoked an investigation into his wonderful abilities. "Learned professors were convinced," wrote Milbourne Christopher (1970, 46), "that Hans could work out his own solutions to mathematical problems and had a better knowledge of world affairs than most fourteen-year-old children." However, psychologist Oskar Pfungst soon determined that questioners—including Hans's trainer—were providing unintentional cueing. Pfungst discovered that Hans began stamping when the questioner leaned forward to observe the horse's hoof and stopped only when that person relaxed after the correct number was given. Pfungst even played the role of Hans by rapping with his hand while friends posed questions. Of 25 questioners, all but 2 gave the beginning and ending cues without being aware of doing so (Christopher 1970; Sebeok 1986).

Of course, trainers could deliberately cue their animals and practice other deceptions, such as secretly gleaning information that the animal would then reveal "psychically." In 1929, the man who later coined the term ESP, Dr. J. B. Rhine, was taken in by a supposedly telepathic horse named Lady Wonder. Rhine believed that Lady actually had psychic power, and he set up a tent near her Virginia barn so that he could scientifically study her apparent abilities. Lady was trained to operate a contraption—somewhat like an enlarged typewriter—consisting of an arrangement of levers that activated alphabet cards. Lady would sway her head over the levers, then nudge one at a time with her nose to spell out answers to queries (Christopher 1970; Jay 1986).

Magician Milbourne Christopher (1970) had an opportunity to assess Lady's talents on a visit in 1956. As a test, Christopher gave Lady's trainer, Mrs. Claudia Fonda, a false name, "John Banks." (The real Banks had exhibited the "talking" horse, Morocco, mentioned earlier.) When Christopher subsequently inquired of Lady, "What is my name?," the mare obligingly nudged the levers to spell out B-A-N-K-S.

Another test involved writing down numbers which Lady then divined. When he was given a narrow pad and a long pencil, Christopher suspected that Mrs. Fonda might be using a professional mentalists' technique known as *pencil reading*, which involves subtly observing the movements of the pencil to learn what is being written. Therefore, he pretended to write a bold "9"—but while going through the motions, he only touched the paper on the downstroke, producing a "1." Although he concentrated on the latter number, Lady indicated that the answer was 9.

In short, as the noted magician and paranormal investigator observed, Mrs. Fonda gave a "slight movement" of her training rod whenever Lady's head was at the correct letter. That was enough to cue the swaying mare to stop and nudge that lever. Thus, Lady was revealed to be a well-trained animal, not a telepathic one (Christopher 1970, 39–54; Nickell 1989, 9–12). No doubt the same was true of her predecessors, whose exhibitors were often performing magicians.

In one case a "talking" animal was allegedly just that: a mongoose that spoke in complete sentences. Gef, as he was called, spoke not only English but also many foreign phrases. He appeared in 1931 on the Irving farm on the Isle of Man (in the Irish Sea), but was never reliably seen. Instead, he tossed stones at unwelcome visitors, and "urinated" through cracks in walls. Although he was partial to the family's 12-year-old daughter, Viorrey, and allegedly lived in her room, he sometimes mischievously locked her inside with a lock that reportedly could only be accessed from outside the room. Psychic investigators supposed that Gef was a poltergeist or perhaps a ghost.

Not surprisingly, there were skeptics, including many fellow residents on the Isle of Man, who believed that Viorrey was playing pranks. They accused her of using ventriloquism and other tricks, the effects of which were hyped by family members, reporters in search of a story, and credulous paranormalists. In fact, a reporter for the *Isle of Man Examiner* once caught Viorrey making a squeaking noise, although her father insisted that the sound had come from elsewhere in the room (*Psychic Pets* 1996, 72–83). In part the case recalls the celebrated magician/ ventriloquist Signor Antonio Blitz, who enjoyed strolling through a village and engaging in conversation with horses tied at hitching posts. Reportedly, he also "once discussed the state of the weather with a dead mackerel in a fish market and almost created a panic" (Christopher 1970, 49).

Psychic Pets

Trickery aside, what about reports of apparent animal ESP? Anecdotal evidence suggests that some animals may have precognitive awareness of various types of natural catastrophes, becoming agitated before earthquakes, volcanic eruptions, cyclones, and other events. However, the creatures may actually be responding to subtle sensory factors—such as variations in air pressure and tremors in the ground—that are beyond the range of human perception (Guiley 1991, 22–25).

Something of the sort may explain some instances of apparent animal prescience. For example, a Kentucky friend of mine insists that his dogs seem to know when he has decided to go hunting, exhibiting marked excitement even though they are lodged some distance away from the house. However, it seems possible that they are either responding to some unintended signal (such as recognizing certain noises associated with his getting ready for a hunting trip) or that he is selectively remembering those occasions when the dogs' excitement happens to coincide with his intentions. Another friend says he once had dogs who seemed to know when he was going to take them for a walk, but he later realized that he must have unconsciously signaled them (such as by glancing in the direction of their hanging leashes).

There is also considerable anecdotal evidence of animals supposedly knowing when their masters were about to suffer harm or were being harmed (Guiley 1991). The operable word here is *anecdotal*: such tales are notoriously untrustworthy. For example, they may be subject to selective recall, so that after a death, say, the deceased's dog is recalled to have "acted strangely" sometime before; other instances of the animals' odd behavior, that did not coincide with the event, are conveniently forgotten. Other problems with anecdotal evidence include the narrator's ego and bias, memory distortion, and other factors.

Scientific tests of animal "psi" (a parapsychological term applied to ESP and psychokinesis) remain controversial (Ostrander and Schroeder 1971, 134–45; Guiley 1991). Rigorous experimental protocols designed to exclude normal explanations (such as sensory cueing) tend not to show evidence for psi. An example is the report on animals' powers of detection by Wiseman, Smith, and Milton, published in the *British Journal of Psychology* in 1998.

The researchers responded to a suggestion by Rupert Sheldrake that just such a study be undertaken. It followed a formal test of the alleged phenomenon by an Austrian television company, which test focused on an English woman and her dog and seemed successful. Wiseman et al. (1998) conducted four experiments designed to rule out the pet's responding to routine or picking up sensory cues (either from the returning owner or from others aware of the expected time of return), as well as people's selective memories, selective matching, and other possible normal explanations.

In all four experiments, the dog failed to detect accurately when her owner set off for home, thus contradicting claims made on the basis of the previous Austrian television study. The experiments suggested "that selective memory, multiple guesses and selective matching could often have sufficient scope to give an owner the impression of a paranormal effect."

Pet Psychics

People who are both devoted to their pets and credulous about the paranormal may easily fall prey to unsubstantiated claims made by pet psychics. Some profess to treat animals' emotional problems, for example, after supposedly communicating with them by ESP or other paranormal means, such as astrology or assistance from the seer's "spirit guides" (MacDougall 1983, 532; Cooper and Noble 1996, 97–113).

After studying pet psychics at work—including Gerri Leigh (with whom I appeared on *Springer*) and Sonya Fitzpatrick (star of Animal Planet's *The Pet Psychic*)—I find that they impress audiences with some very simple ploys. Consciously or not, they are essentially using the same fortunetellers' technique of cold reading that is used with human subjects. This is an artful method of gleaning information from someone while giving the impression that it has been obtained mystically (Hyman 1977). After all, it is the pet owners, not the pets themselves, who "validate" the pronouncements. Here is a look at some of the common cold-reading techniques used by pet psychics.

1. *Noting the obvious.* Fitzpatrick (2002) visits an animal clinic with a couple and their infant daughter to tell them which dog is right for their family. After the selection is narrowed to three choices, each is brought out in turn. The first is ambivalent; the second ignores everyone; and the third, Patty, greets the couple and nuzzles the child. Sonya writes her choice on a slip of paper and it proves to be the same the couple made: Patty. The audience applauds; Patty was apparently their choice too! (I know she was mine!)
2. *Making safe statements.* Fitzpatrick (2002) announces that one pooch "says" he wants to go out more often, and the dog's owners accept the assertion. Similarly, Gerri Leigh (1992) tells the owner of an outgoing little dog, which immediately licks Leigh's hand, that the animal "fears no one"—but then she quickly adds that it is "not an unconditional lover." She continues by stating that the pet is "independent" and "not a yes dog." Such virtually universal declarations are not apt to be challenged.
3. *Asking questions.* Psychics frequently seem to be providing information when in fact they are fishing for it. The asking of a question may, if it is correct, allow the reader to be credited with a hit; otherwise it will seem an innocent query. For instance, Fitzpatrick (2002) asks a dog owner, "When was there someone who was with him who went away?" (Unfortunately, this is too good a hit, since the young woman seems puzzled and replies that it could have been various persons—possibly, one imagines, former boyfriends or other acquaintances.) Questioning also keeps the reader from proceeding too far down a wrong path and allows for mid-course corrections.
4. *Offering vague statements that most people can apply specifically to themselves.* Alleged psychics take advantage of what is known as "the Barnum effect"—named after showman P. T. Barnum who strove to provide something for everyone (French et al. 1991). They learn that people will respond to a vague, generalized state-

ment by trying to fit it to their own situations. Thus Fitzpatrick (2002) tells the owner of a pet iguana that the creature had experienced "a move." Now, most people can associate a "move" with their pet: when they acquired it, when they changed residences, when they left it with someone to go on vacation, and so on. Thus the pet psychic was credited with a hit (never mind that she incorrectly referred to the female iguana as "he").

5. *Returning messages to animals.* People who are convinced that pets give information to psychics may be willing to believe the reverse. Thus Fitzpatrick (2002) claims to give animals "messages"—for example, a clarification of something by the owner—by silently concentrating for a moment.

These and other techniques help convince the credulous that pet psychics have telepathic or clairvoyant or other powers. Some, like New York psychic Christa Carl, even claim to use these powers to help locate lost pets. Carl gained notoriety "for being called in to find Tabitha, the cat who disappeared on a Tower Air flight." Actually, my interpretation of the case is that Carl did not find the cat, but that the cat found Carl—or rather, found her owner. Tabitha was known to be hiding on the airplane; after 12 days and 30,000 miles of flight that engendered a large amount of negative publicity and a threatened lawsuit, the airline grounded the plane so the animal could be retrieved. The cat eventually came out to her owner—and to Carl, who claimed the credit for supposedly helping the animal resolve a problem with "one of her past lives" and "showing her how to come out" of the plane's drop ceiling (Cooper and Noble 1996).

To find other lost animals, Carl claims that she uses "visualization" to help them "find their way home." Thus, if an animal returns, Carl can claim credit. If not, she has a ready rationalization: Some animals do not wish to come back and, says Carl, "I have to respect the animal's wishes" (Cooper and Noble 1996).

Some pet psychics offer still other services. For example, Oklahoma "equine parapsychologist" Karen Hamel-Noble claims to heal horses. She uses her hands to detect "the source of weakness in their energy fields"—that is, their imagined auras—and then supplies compensating "energy" from herself (Cooper and Noble 1996). However, auras remain scientifically unproven, and tests of psychics' abilities to see them have repeatedly failed (Nickell 2000); Hamel-Noble's claims require proof, not just her statements of feelings and other assertions. Perhaps the animals' perceived recoveries from illnesses are merely response to their natural healing mechanisms and the medical treatments Hamel-Noble provides them—including penicillin injections (Cooper and Noble 1996).

Pet Mediums

In the popular imagination, animals, like their human counterparts, may continue their existence after death. There are many reports of animal apparitions. Because pets are loved and often regarded as members of a family, it is not surprising that people occasionally experience "visitations" from their departed animal friends just as they do from their human ones. However, these seem to have explanations similar to those of other apparitional experiences. For example, some who hear a dog's phantom bark or footsteps, or see (as one reported) "a shadow jump up on the bed," may do so just after rousing from sleep (Cohen 1984, 137–49) and may thus be having *waking dreams*. These common hallucinations occur in the twilight between being awake and asleep and exhibit content that "may be related to the dreamer's current concerns" (Baker 1990, 179–82). Similarly, apparitions that are seen during wakefulness tend to occur when the perceiver is tired, daydreaming (perhaps while performing routine work), or in a similar state or situation (Nickell 2001a, 291–92).

With the advent of spiritualism—the belief that the dead can be contacted—certain self-styled "mediums" began to offer themselves as intermediaries with the spirit realm. Some produced bogus spirit "materializations" and other physical phenomena, but these were frequently exposed as tricks by investigators such as magician Harry Houdini. Today's mediums tend to limit themselves to purely "mental phenomena," that is, the use of "psychic ability" to obtain messages from "the other side."

Today's mediums—including James Van Praagh, John Edward, Rosemary Altea, George Anderson, and Sylvia Browne—appear to rely largely on the old psychics' standby, cold reading. In fact, Edward (whose real name is John MaGee, Jr.) came to mediumship after a stint as an erstwhile fortuneteller at psychic fairs; now, however, he styles himself a "psychic medium." On *Dateline NBC,* he was caught cheating, attempting to pass off some previously gained knowledge as spirit revelation (Nickell 2001b).

Mediums like Edward and Van Praagh occasionally mention a pet—usually a dog—in a reading. Given the Barnum effect (discussed earlier), this usually gets a hit. For instance, on the television show *Larry King Live* (26 February1999), Van Praagh told a caller: "I'm also picking up something on a dog. So I don't know why, but I'm picking up a dog around you." Note the vagueness of the reference: there is not even an indication of whether the animal is dead or alive or what link it might have to the person. But the caller offers the validation, "Oh, my dog died two years ago."

Some pet psychics, like Christa Carl, conduct "séance readings" for animals who have "passed over." When asked to give an example of such a séance, she replied (in Cooper and Noble 1996, 102):

> Brandy, a dog, had been placed in a kennel by her owner when she got married. She broke away from the kennel and got killed.
>
> Her owner called me and told me she was having a hard time and wanted to communicate with Brandy. When

> I did the reading with Brandy, I learned from her that she didn't know why she had been put in the kennel. She had felt abandoned, unloved, uncared for.
>
> Her owner should have told her ahead of time why she needed to put her in a kennel. I explained it to Brandy, and now she's at peace.

Of course, there is not the slightest bit of evidence that the spirit was contacted or that, in fact, it existed anywhere other than in the imaginations of Christa Carl and, of course, the dog's grieving, guilt-ridden, and credulous owner.

This pinpoints the inevitable problem with claims involving psychic pets and pet psychics. They are based on anecdotal evidence—wonderful tales of psychic and mediumistic success—but are not supported by scientific investigation.

REFERENCES

Baker, Robert A. 1990. *They Call It Hypnosis*. Buffalo, N.Y.: Prometheus Books.

Christopher, Milbourne. 1962. *Panorama of Magic*. New York: Dover.

———. 1970. *ESP, Seers & Psychics*. New York: Thomas Y. Crowell.

Cohen, Daniel. 1984. *The Encyclopedia of Ghosts*. New York: Dorset Press.

Cooper, Paulette, and Paul Noble. 1996. *100 Top Psychics in America*. New York: Pocket Books.

Fitzpatrick, Sonya. 2002. "The pet psychic." Animal Planet television series, aired 7 March.

French, Christopher C., et al. 1991. Belief in astrology: A test of the Barnum effect. *Skeptical Inquirer* 15, no. 2 (Winter): 166–72.

Guiley, Rosemary Ellen. 1991. *Harper's Encyclopedia of Mystical & Paranormal Experience*. New York: HarperCollins.

Hyman, Ray. 1977. Cold reading: How to convince strangers that you know all about them. *Skeptical Inquirer* 1, no. 2 (Spring/Summer): 18–37.

Jay, Ricky. 1986. *Learned Pigs & Fireproof Women*. London: Robert Hale.

Leigh, Gerri. 1992. Interview/appearance on *The Jerry Springer Show*, 16 March.

MacDougall, Curtis D. 1983. *Superstition and the Press*. Buffalo, N.Y.: Prometheus Books.

Nickell, Joe. 1989. *The Magic Detectives.* Buffalo, N.Y.: Prometheus Books.
———. 2000. Aura photography: A candid shot. *Skeptical Inquirer* 24, no. 3 (May/June): 15–17.
———. 2001a. *Real-life X-Files.* Lexington, Ky.: University Press of Kentucky.
———. 2001b. John Edward: Hustling the bereaved. *Skeptical Inquirer* 25, no. 6 (November/December): 19–22.
Ostrander, Sheila, and Lynn Schroeder. 1971. *Psychic Discoveries Behind the Iron Curtain.* New York: Bantam.
Psychic Pets and Spirit Animals: True Stories from the Files of Fate Magazine. 1996. New York: Gramercy Books.
Sebeok, Thomas A. 1986. Clever Hans redivivus. *Skeptical Inquirer* 10, no. 4 (Summer): 314–18.
Wiseman, Richard, Matthew Smith, and Julie Milton. 1998. Can animals detect when their owners are returning home? An experimental test of the "psychic pet" phenomenon. *British Journal of Psychology* 89: 453–62.

34

Cryptids "Down Under"

The term *cryptid* was coined to refer to unknown animal species or to those which, though believed extinct, may only have eluded scientific rediscovery (Coleman and Clark 1999, 75). Examples of the former are the yowie (Australia's version of Bigfoot) and the bunyip (a swamp-dwelling, hairy creature with a horselike head) (Coleman and Clark 1999, 49–50, 255–57). An example of the latter is the thylacine.

At a skeptic's convention in Sydney in 2000, Australian paleontologist Mike Archer discussed the thylacine as part of his talk, "Creationism and Its Negative Impact on Good Science." Also known as the Tasmanian tiger, the *thylacinus cynocephalus* was a wolflike marsupial with prominent stripes on its back (Figure 34-1). It became extinct on the mainland some 2,500 years ago, but continued to exist on Tasmania (one of a group of islands comprising Australia's smallest state), until it finally succumbed to habitat destruction, bounty hunters, and other

FIGURE 34-1. **Thylacine or "Tasmanian tiger"—believed extinct since 1936—as a mounted specimen in the Australian Museum. (Photograph by Joe Nickell.)**

forces. The last known thylacine died in a zoo in 1936 (Park 1985). Nevertheless, since then hundreds of sightings have been reported, some by multiple eyewitnesses. Alleged sightings were on the increase even in the 1980s. However, there were few reports of attacks on sheep or other domestic animals, as would have been expected if thylacines were making a comeback (Park 1985), so the increase in sightings might have been due to the bandwagon effect. As with reported sightings of other cryptids, the tendency to see what one expects to see is powerful. Paranormalist Rupert T. Gould called this tendency "expectant attention" (Binns 1984, 77–78).

Although Tasmania would seem the most credible locale, the thylacine has allegedly also been sighted often on the mainland—albeit in relatively isolated areas (Coleman and Clark 1999, 239). For example, thylacines are "frequently reported seen in the coastal border country between Victoria and South Australia" (Gilroy 1995, 74). Indeed, in November 2000, as Australian skeptics Bob Nixon, Richard Cadena, and I drove along the Great Ocean Road from Melbourne to Warrnambool, Bob recalled one reported Tasmanian tiger sighting some years ago near Lorne (where we ate lunch). This was an area of virgin "bush" country (a eucalypt forest), but, alas, all we saw was beautiful scenery. I also kept an eye out for the thylacine while looking for the yowie—to be discussed presently—in the Blue Mountains, another area where the striped creature is reportedly seen (Gilroy 1995).

Hope springs eternal, but it increasingly appears that if the thylacine is not to remain elusive forever, an idea of paleontologist Mike Archer's must prevail. Archer, who is also director of the Australian Museum, has suggested resurrecting the species. Using DNA from a preserved specimen, he proposes to clone the creature, giving us a glimpse of that possibility at the skeptics conference. (For a discussion of the relevant biotechnology, see Lanza et al. 2000.)

The yowie, in contrast, has left only meager traces of its supposed existence, like those of other hairy man-beasts reported around the world. These include the Himalayan yeti, the North American sasquatch, and similar creatures alleged to inhabit remote regions of China, Russia, southeast Asia, and elsewhere.

The yowie is a fearsome, hairy creature of Aboriginal mythology. Also called Doolagahl ("great hairy man"), it is venerated as a sacred being from the time of creation, which the Aborigines call the Dreamtime. An alleged sighting by a hunting party of settlers in 1795 was followed by increased reports from the mountainous regions of New South Wales in the nineteenth century. For example, in 1875 a coal miner exploring in the Blue Mountains west of Sydney reportedly stalked a hairy, apelike animal for a distance before it finally eluded him. Sightings of the yowie mounted as settlers penetrated the country's vast interior, and yowie hunter Rex Gilroy (1995, 197) notes that his files now "bulge with stories from every state."

The self-described "'father' of yowie research," Gilroy (1995, 202) boasts the acquisition of some 5,000 reports, together with a collection of footprint casts, but he complains of "a lifetime of ridicule from both ignorant laymen and scientists alike." When Peter Rodgers and I ventured into the Blue Mountains, we experienced something of the prevalent local skepticism at the information center at Echo Point (in the township of Katoomba). Staffers there were emphatic that the yowie was a mythical creature pursued by a few fringe enthusiasts. (To them, yowies exist only as popular toys and chocolate figures marketed by Cadbury.)

Nevertheless, to Gilroy "the Blue Mountains continues to be a hotbed of yowie man-beast activities—a vast region of hundreds of square miles still containing inaccessible forest regions seldom if ever visited by Europeans." The fabled creatures are known there, he says, as the "Hairy Giants of Katoomba" and also as the "Killer Man-Apes of the Blue Mountains" (Gilroy 1995, 212).

In the Katoomba bushland, Peter and I took the celebrated "steepest incline railway in the world" (built as a coal-mine transport in 1878) down into Jamison Valley. The miserable weather gave added emphasis to the term *rainforest,* through which we "bushwalked" (hiked) west along a trail. We passed some abandoned coal mines, which Peter humorously dubbed "yowie caves," before eventually retracing our route. We saw no "Hairy Giants of Katoomba" but, to be fair, we encountered little wildlife at all. The ringing notes of the bellbird did herald our visit and announce that we were not alone.

Resuming our drive, we next stopped at Meadlow Bath, an historic resort area. From the "haunted" Hydro Majestic Hotel overlooking the Megalong Valley—also reputed to be yowie country (Gilroy 1995, 217–18)—we surveyed a countryside that was largely shrouded in fog. Proceeding through Blackheath and Victoria Pass (where a bridge is said to be haunted by a female specter [Davis 1998, 95–97]), we continued on to Hartley, then took a narrow, winding road some 44 kilometers to Jenolan Caves. Gilroy (1995, 219) states that the Aborigines believed the caves were used in ancient times as yowie lairs, and he cites reported sightings and discoveries of footprints in the region. (For millennia the Jenolan area was known to the local Aborigines as *Binoomea*, meaning "holes in the hill." According to legend, the first non-Aborigine to discover the area was a bushranger, an escaped convict named McKeown, who used it as a refuge in the 1830s. Once, after a pursuer had followed him for miles, he disappeared, but his tracks "led up to a wild cavern and into it . . . and burst again into open day, and the route lay along a rugged gorge for some three miles" [Bates 2000, 23].)

Except for passing through the Grand Arch, a majestic limestone-cavern entranceway into a hidden valley, and surveying the spectacular grotto called Devil's Coachhouse, we avoided the caves themselves in order to continue our cryptozoological pursuit. (This despite the discovery therein of a skeleton of the extinct thylacine [*Gregory's* 1999].) We instead searched the surrounding mountainous terrain (see Figure 34-2) for signs of the elusive yowie, again without success. Here and there the raucous laughter of the kookaburra seemed to mock our attempt. Neither did we encounter another claimed paranormal entity—a ghostly lady—when we dined at the "haunted" Jenolan Caves House. An employee told us he had worked at the site for three years without seeing either a yowie or the inn's resident "ghost," and he indicated that he believed in neither.

Failing to encounter our quarry, we ended our hunt relatively unscathed—soaked, to be sure, and I with a slightly wrenched knee. But consider what might have been: headlines screaming, "Skeptics mauled by legendary beast!"—a tragic way to succeed, certainly, and with no guarantee, even if we survived, that we would be believed! Even Gilroy conceded (1995, 202) that "nothing short of actual physical proof—such as fossil or recent skeletal remains or a living specimen—will ever convince the scientific community of the existence of the 'hairy man.'"

That, however, is as it should be: In many instances the touted evidence for Bigfoot-type creatures—mostly alleged sightings and occasional footprints—has been shown to be the product of error or outright deception (Nickell 1995, 222–31). Cryptozoologists risk being thought naïve when they too quickly accept the evidence of "manimal"

FIGURE 34-2. **Terrain of the legendary yowie (Australia's Bigfoot) viewed through Carlotta Arch in the Jenolan Caves region. (Photograph by Joe Nickell.)**

footprints. "Some of these tracks," insists Gilroy (1995, 224), "have been found in virtually inaccessible forest regions by sheer chance and, in my view, must therefore be accepted as authentic yowie footprints." It seems not to have occurred to the credulous monsterologist that a given "discoverer" might actually be the very hoaxer. Thus, the debate continues.

REFERENCES

Bates, Geoff. 2000. Historic Jenolan Caves. In *Blue Mountains Tourist*, Olympic ed. (citing *Government Gazette*, 19 August 1884).

Binns, Ronald. 1984. *The Loch Ness Mystery Solved.* Buffalo, N.Y.: Prometheus Books.

Coleman, Loren, and Jerome Clark. 1999. *Cryptozoology A to Z.* New York: Fireside (Simon & Schuster).

Davis, Richard. 1998. *The Ghost Guide to Australia.* Sydney: Bantam Books.

Gilroy, Rex. 1995. *Mysterious Australia.* Mapleton, Queensland, Australia: Nexus Publishing.

Gregory's Blue Mountains in Your Pocket. 1999. 1st ed. Map 238. Macquarie Centre, N.S.W.: Gregory's Publishing.

Lanza, Robert P., et al. 2000. Cloning Noah's Ark. *Scientific American*, November, 84–89.

Nickell, Joe. 1995. *Entities: Angels, Spirits, Demons, and Other Alien Beings.* Amherst, N.Y.: Prometheus Books.

Park, Andy. 1985. Is this toothy relic still on the prowl in Tasmania's wilds? *Smithsonian*, August, 117–30.

35

Joseph Smith

A Matter of Visions

Past attempts to understand the motivations of visionaries, psychics, faith healers, and other mystics—seers like Mormon founder Joseph Smith—have often focused on a single, difficult question: Were they mentally ill, or were they instead charlatans? Increasingly, there is evidence that this may be a false dichotomy, that many of the most celebrated mystics may in fact simply have possessed fantasy-prone personalities. Called "fantasizers," such individuals fall within the normal range and represent an estimated 4 percent of the population.

This personality type was characterized in 1983 in a pioneering study by Sheryl C. Wilson and Theodore X. Barber. Some 13 shared traits were identified:

1. being susceptible to hypnosis
2. having imaginary companions in childhood

3. fantasizing frequently as a child
4. adopting a fantasy identity
5. experiencing imagined sensations as real
6. having particularly vivid sensory experiences
7. reliving (not just recalling) experiences
8. claiming psychic powers
9. having out-of-body experiences
10. receiving special messages from spirits, higher intelligences, or the like
11. having healing powers
12. encountering apparitions
13. experiencing hypnagogic or hypnopompic hallucinations ("waking dreams") with classical imagery (such as spirits or monsters from outer space).

As in previous studies (Nickell 1997), I consider the presence of six or more of these traits in an individual to be indicative of fantasy-proneness. (Anyone may have a few of these traits, and only the very rare person would exhibit all of them.)

Wilson and Barber also found evidence suggesting that "individuals manifesting the fantasy-prone syndrome may have been overrepresented among famous mediums, psychics, and religious visionaries of the past" (1983, 371). These researchers further found that biographies could yield evidence that a subject was a fantasizer, and they reached such a determination in the case of Mary Baker Eddy, founder of Christian Science; Joan of Arc, the Catholic saint; and Gladys Osborne Leonard, the British spiritualist, among others. It should be noted that Wilson and Barber also included Theosophy founder Madame Helena P. Blavatsky, although her propensity for trickery during séances is well known. Deception and fantasy are obviously not mutually exclusive, as we shall see in the case of Joseph Smith himself, whom most non-Mormon scholars regard as a veritable confidence man but whom Wilson and Barber also specifically include in their list of historical fantasizers (372).

Joseph Smith, Jr. (1805–1844) was the founder of the Church of Jesus Christ of Latter-day Saints, popularly known as the Mormon church. He was born 23 December 1805, in Sharon, Vermont, the third of nine children of Joseph and Lucy (Mack) Smith. A poor, unchurched, but religious family, the Smiths migrated in 1816 to Palmyra, New York. A contemporary recalled the young Joe as a disheveled boy, dressed in patched clothing with homemade suspenders and a battered hat.

> He was a good talker, and would have made a fine stump speaker if he had had the training. He was known among the young men I associated with as a romancer of the first water. I never knew so ignorant a man as Joe was to have such a fertile imagination. He never could tell a common occurrence in his daily life without embellishing the story with his imagination; yet I remember that he was grieved one day when old Parson Reed told Joe that he was going to hell for his lying habits [quoted in Taves 1984, 16].

At the age of 14, Smith later wrote, he became troubled by the various religious revivals in the area, and so he sought a wooded area where he hoped to commune directly with God. As he later wrote:

> It was the first time in my life that I had made such an attempt, for amidst all my anxieties I had never as yet made the attempt to pray vocally. . . . I kneeled down and began to offer up the desires of my heart to God. I had scarcely done so, when immediately I was seized upon by some power which entirely overcame me, and had such an astonishing influence over me as to bind my tongue so that I could not speak. Thick darkness gathered around me, and

> it seemed to me for a time as if I were doomed to sudden destruction. But, exerting all my powers to call upon God to deliver me out of the power of this enemy which had seized upon me, and at the very moment when I was ready to sink into despair and abandon myself to destruction—not to an imaginary ruin, but to the power of some actual being from the unseen world, who had such marvelous power as I had never before felt in any being—just at this moment of great alarm, I saw a pillar of light exactly over my head, above the brightness of the sun, which descended gradually until it fell upon me.

Smith continued:

> It no sooner appeared than I found myself delivered from the enemy which held me bound. When the light rested upon me I saw two personages, whose brightness and glory defy all description, standing above me in the air. One of them spake unto me, calling me by name, and said—pointing to the other—"*This is my beloved Son, hear Him.*"
>
> My object in going to inquire of the Lord was to know which of all the sects was right, that I might know which to join. No sooner, therefore, did I get possession of myself, so as to be able to speak, that I asked the personages who stood above me in the light, which of all sects was right—and which I should join. I was answered that I must join none of them, for they were all wrong, and the personage who addressed me said that all their creeds were an abomination in His sight: that those professors were all corrupt; that "they draw near to me with their lips, but their hearts are far from me; they teach for doctrines the commandments of men: having a form of godliness, but

> they deny the power thereof." He again forbade me to join with any of them: and many other things did he say unto me, which I cannot write at this time. When I came to myself again, I found myself lying on my back, looking up into heaven [quoted in Brodie 1993, 21–22].

Although Smith gave different versions of his visions (Persuitte 2000), his biographer Fawn M. Brodie notes that somewhat similar experiences "were common in the folklore of the area"—an indication that Joseph's experience was probably genuinely real to him. A few years later, at the age of 17, he had another experience. Although again there are different versions, it is described in terms entirely consistent with an actual hypnogogic hallucination (waking dream):

> A personage appeared at my bedside, standing in the air. . . . He had on a loose robe of most exquisite whiteness. . . . His whole person was glorious beyond description. I was afraid; but the fear soon left me. He called me by name, and said unto me that he was a messenger sent from the presence of God to me and that his name was Moroni; that God had a work for me to do; and that my name should be had for good and evil among all nations, kindreds and tongues [quoted in Taves 1984, 277].

Then Smith received the crucial communication. Moroni supposedly told him where to find a book, written on gold plates, that gave "an account of the former inhabitants of this continent," together with two stones—the biblical Urim and Thummin.

It should be mentioned that during this period the young Smith was engaged in "money-digging," searching for hidden treasure by scrying (i.e., crystal gazing) or by dowsing (using a witch-hazel wand or mineral rod that was supposedly attracted by whatever was sought).

Some have seen this as a form of fraud, but Taves (1984, 19) points out that the practice was an old one and that treasure-laden burial mounds dotted the area. Nevertheless, Joseph Smith, Jr., was arrested on the complaint of a neighbor that he was "a disorderly person and an impostor." Witnesses, who were divided as to the genuineness of Joseph's skill, reported that he looked at a special stone which he placed in his hat. The dispensation of the case is unclear, but apparently Smith agreed to leave town (Persuitte 2000, 40–53; Taves 1984, 17–18).

Brodie describes the young treasure-seeker as having "an extraordinary capacity for fantasy," which, she says, "with proper training might even have turned him to novel-writing." She also says that "[h]is imagination spilled over like a spring freshet. When he stared into his crystal and saw gold in every odd-shaped hill, he was escaping from the drudgery of farm labor into a glorious opulence." She adds: "Had he been able to continue his schooling, subjecting his plastic fancy and tremendous dramatic talent to discipline and molding, his life might never have taken the exotic turn it did" (Brodie 1993, 27).

Joseph supposedly followed Moroni's directions and discovered, on what is now known as the Hill Cumorah, the golden book and other items in a stone box. Or so he alleged. He was now convinced that he had been divinely chosen as the instrument by which the "corrupted" Bible was to be restored. Smith "translated" the text (or, more likely, he imagined it, while borrowing from certain contemporary writings). His bride, Emma, was his first scribe, followed by an early convert named Martin Harris. To effect the translation, Smith sat on one side of the room staring into his peepstone (in the type of reverie practiced by scryers), with Harris on the other side writing at a table and a blanket across a rope separating the two (Taves 1984, 35–40).

Like Edgar Cayce, the "sleeping prophet," Smith sometimes temporarily lost his gift of seeing—notably as happened after Harris managed to lose the first 116 pages of the manuscript, leaving the prophet inconsolable for a time. The same waning of power occurred with regard to Smith's ability to perform healings. Although he cast out a devil from one man and healed a woman of a "rheumatic" arm (Taves 1984, 63–69), on another occasion he tried unsuccessfully to heal the sick and even, in one instance, to revive the dead (Taves 1984, 70).

The *Book of Mormon* was published in 1830. Shortly afterward, Joseph Smith and his associate Oliver Cowdery, having been conferred priests by divine revelation to Smith, officially founded the Church of Christ at Fayette, New York. Eight years later the name was changed to the Church of Jesus Christ of Latter-day Saints (Hansen 1995, 365).

An invitation from Sidney Rigdon, onetime associate of revivalist Andrew Campbell, led Smith and his New York brethren to found a Mormon settlement at Kirtland, Ohio. There, Smith claimed to experience a further series of revelations (published in 1833 under the title *Book of Commandments*), which expanded his theological principles. The revelations directed the Saints to gather into communities in a patriarchal order and to erect a temple at the center of the community. In 1838, financial problems caused abandonment of Kirtland.

Smith also founded communities in Missouri, but in 1839 he and his followers were driven ruthlessly from that state by anti-Mormon vigilantes. The Saints then gathered at a settlement called Nauvoo on the Mississippi River. By 1844, it was the most populous city in Illinois and entirely under Mormon control (Hansen 1995, 365).

It was at Nauvoo that Joseph Smith met his end. He had increasingly acted on pretensions of grandeur that led him to become leader of the Mormon militia, bedecked in the uniform of a lieutenant general, and an announced candidate for the United States presidency. As before, mobs of anti-Mormons plagued Smith and his followers, and when the latter destroyed an anti-Mormon press, Smith was jailed on a charge of riot. On 27 June1844, a mob stormed the jail, killing Smith and his brother Hyrum (Hansen 1995, 365). Concludes Taves (1984, 213): "It was over. The gangly, ill-clad youth who had regaled his Palmyra neighbors with fanciful tales had come a long, long way before reaching the end of his road. Others would continue what he had started."

As this brief sketch illustrates, Joseph Smith, Jr., had numerous traits collectively indicating that his was a fantasy-prone personality. The traits include:

1. easily undergoing self-hypnosis (as during his scrying and translating)
2. frequently fantasizing as a child
3. having imagined sensations that seemed real
4. believing he had divinatory powers
5. receiving special messages from on high
6. believing he had healing powers
7. encountering apparitions
8. experiencing waking dreams with classical imagery.

These eight characteristics—possibly among others—confirm Wilson and Barber's earlier diagnosis and thereby reveal Smith to be typical of religious visionaries who share the characteristics of fantasy-proneness.

REFERENCES

Brodie, Fawn M. 1993. *No Man Knows My History: The Life of Joseph Smith.* 2d ed. New York: Alfred A. Knopf.

Hansen, Klaus J. 1995. "Joseph Smith." In *The Encyclopedia of Religion.* Vols. 13–14. New York: Simon & Schuster.

Nickell, Joe. 1997. The two: A fantasy-assessment biography. *Skeptical Inquirer* 21, no. 4 (July/August): 18–19.

Persuitte, David. 2000. *Joseph Smith and the Origins of the* Book of Mormon. 2d ed. Jefferson, N.C.: McFarland.

Taves, Ernest H. 1984. *Trouble Enough: Joseph Smith and the Book of Mormon.* Buffalo, N.Y.: Prometheus Books.

Wilson, Sheryl C., and Theodore X. Barber. 1983. The fantasy-prone personality. In *Imagery: Current Theory, Research, and Application,* edited by Anees A. Sheikh, 340–387. New York: John Wiley & Sons.

36

In Search of Fisher's Ghost

During an investigative tour Down Under, I was able to examine the persistent legend of "Australia's most famous ghost" (Davis 1998, 16). I was generously assisted by magic historian Peter Rodgers, with whom I shared several other adventures (Nickell 2001).

One writer has commented, "It is a mystery why some ghost stories catch the public's imagination and survive while others, often more shocking and more credible, are forgotten" (Davis 1998, 16–18). Davis cites the story about Frederick Fisher, which has been related in countless newspaper articles, as well as poems, songs, books, plays, an opera, and other venues (Davis 1998) and provided the inspiration for a movie (Fowler 1991). It once attracted the attention of notables like Charles Dickens, who published a version of the story in his magazine *Household Words;* and entertainer John Pepper, who used it as the subject of one of his "Pepper's ghost" stage illusions in Sydney circa 1879 ("Illusionist" 1984). Today, Fisher's ghost remains the subject of an annual festival. All this—even though the ghost reportedly appeared "to just one man on one occasion" long ago (Davis 1998).

The story began on 17 June 1826, with the disappearance of Frederick Fisher. Fisher was a "ticket-of-leave man"—a paroled convict—who had acquired land at Campbelltown and built a shack thereon. Unfortunately, he also caroused there with itinerants and other ticket-of-leave men, including his neighbor and best friend George Worrell (or Worrall). When Fisher found himself in debt and facing possible arrest, he trustingly signed his property over to Worrell, either to conceal or to protect his assets. When Fisher was released from prison, after a six-month stretch, he returned to his farm only to find that Worrell had been claiming it as his own.

After Fisher disappeared, Worrell resumed possession of the property, telling anyone who inquired that his friend had returned to England in search of his estranged family. The facts that Worrell wore Fisher's clothes and offered a crudely forged receipt to prove his ownership of one of Fisher's horses soon raised suspicions.

On 23 September, the Colonial Secretary's Office offered a reward for "the discovery of the body" of Frederick Fisher, or a lesser reward for proof that he had "quitted the Colony" ("Supposed Murder" 1826). Subsequently, a local man named James Farley reportedly had an encounter with the ghost of Fisher. Farley was walking near Fisher's property one night and saw an apparition of the missing man sitting on a fence, glowing eerily and dripping blood from a gashed head. Moaning, the phantasm "pointed a bony finger in the direction of the creek that flowed behind Fisher's farm" (Davis 1998) (see Figure 36-1). Thus prompted to search the area, police soon dug up Fisher's corpse. Worrell was convicted of the murder and reportedly confessed just before his hanging (Fowler 1991, 13).

Such are the main outlines of the story. Queensland writer Richard Davis observes, in his *Ghost Guide to Australia* (1998), "From the begin-

FIGURE 36-1. **An artist's impression of the appearance of Fisher's ghost beckoning to a resident named Farley, in 1826.**

ning distortions occurred—almost every aspect of the story was changed and romanticised so that truth became indistinguishable from fiction." Indeed, the version published by Charles Dickens ("Fisher's Ghost" 1853) contains numerous altered details—"Penrith" for *Campbelltown*, "Smith" for *Worrell*, and so on—that link it to a *fictionalized* account written by Australian writer John Lang (n.d.).

Those promoting the tale cite an alleged deathbed statement by the percipient James Farley (or "John Hurley" in the earliest versions [Cranfield 1963]). Queried about the matter on his deathbed, Farley supposedly raised himself on an elbow and told his friend: "I'm a dying man, Mr. Chisholm. I'll speak only the truth. I saw that ghost as plainly as I see you now" (Davis 1998; Cusack 1967, 3). Unfortunately, the story is not only unverified, but also has a suspiciously literary quality about it.

In fairness, it should be acknowledged that debunkers have offered their share of doubtful claims as well. One purported explanation for the ghost was given by a 73-year-old barber. He said he heard it from his grandfather, who in turn allegedly learned it from an ex-convict who had secretly witnessed the murder and burial. Wanting to expose the truth, but afraid of being implicated, he hit on a plan. He fashioned a pair of cloaks—one white, another black—and wore the first at night to simulate ghostliness. When some traveler happened by, he moaned and pointed to the burial site in the swamp. Then, readying the black cloak as he walked toward that spot, he suddenly pulled it over him so that "to the terrified onlooker it seemed that the ghost had suddenly disappeared." Supposedly this repeated ruse brought the desired result, and the corpse was searched for and discovered—believe it or not! ("Ghost" 1955).

Another hand-me-down tale was related by a 74-year-old resident. He said that Farley simply "saw a man whom he took to be Fisher (but it was not Fisher) sitting on the rail of the bridge." When the man "dropped from the rail of the bridge apparently into the weeds" and so seemed to vanish, "Farley thought it must have been a ghost on account of the sudden disappearance" (Lee 1963). Though such an incident *could* happen, there is no good evidence that it *did*.

Not surprisingly, those who are inclined to dismiss ghost stories have suggested that the tale was simply a journalistic invention. One writer stated that "there can be little doubt that it was a hoax first published by a Sydney magazine" (Cranfield 1963). In fact, however, that account—in the *Teggs Monthly* of 1 March 1836—was preceded by an anonymous poem published years earlier in *Hill's Life in New South Wales*. Titled "The Spirit of the Creek," it bore a prefatory note that it was based on the murder of "poor F*****" at Campbelltown. It is important to note that this was a creative production. Not only was *Hills' Life* a literary paper and the narrative written in verse (thus inviting "poetic license"), but the story was actually fictionalized. For example, Fred Fisher became a rich ex-convict named "Fredro" and the murderer Worrell was represented as "Wurlow" (Fowler 1991, 15).

To assess the credibility of the Fisher's ghost story, it is necessary to go back in time, as it were, to the proceedings of the Supreme Criminal Court on 2 February 1827 (Supreme 1827). As others have previously noted (e.g., Cranfield 1963), the trial records make absolutely no mention of a ghost. In addition to this negative evidence, I was struck by the positive evidence in the proceedings that Fisher's missing body had actually been located in a rational rather than supernatural manner. Constable George Looland testified that, on the previous October 20, blood found on several fence rails at the corner of Fisher's paddock led him to search the area. He was assisted by two aboriginal trackers who soon reported traces that they thought was "the fat of a white man" (presumably human tissue) floating on the creek. Proceeding with the search, they came to a spot (apparently identified by a disturbance of the marshy area) that they probed with an iron rod. One of the trackers "called out that there was something there," and a spade was procured to excavate the site. Soon the search party had uncovered the "left hand of a man lying on his side." The coroner was summoned, and (the next morning) the body of Fisher was exhumed and examined, whereupon "several fractures were found in the head" (Supreme 1827).

However the story of Fisher's ghost was actually launched—and it may have originated with the previously mentioned anonymous poem in 1832—the legend has persisted. In the narrative, the phantom behaves as one of those purposeful spirits of yore that sometimes "advised where their bodies might be discovered" (Finucane 1984, 194). Folklorists recognize such tales as types of *supernatural legends*—that is, "supposedly factual accounts of occurrences and experiences which seem to validate superstitions" (Brunvand 1978, 108–9).

Evidence of folklore in progress is quite evident. Numerous variations in the tale (apart from the fictionalizing process) are suggestive of oral transmission. Consider a specific example. Since at least the 1950s, lighthearted vigils for the ghost have been held, with crowds typically gathering at midnight on June 17th. The chosen site is the bridge across Fisher's Ghost Creek because, according to one account, "it was on the rail of the bridge . . . that Fisher's Ghost was always seen" ("Fisher's Ghost" 1957). However, when Peter Rodgers and I made our pilgrimage to the spot, locals told us (and other sources confirmed) that the existing bridge was not in precisely the same place as the original. More significantly, the earliest accounts of the story have the ghost sitting on the rail of a *fence*. With that simple transformation of a *motif* (as folklorists term a narrative element)—from fence rail to bridge rail—the *site* of the purported apparition also became translocated. Nevertheless, "ghost" sightings have been reported there; one of the most noteworthy occurred in 1955, when "a white cow in the distance in the pitch darkness gave some onlookers a scare" ("Fisher's Ghost" 1957).

Clearly, the story of Fisher's ghost has many of the elements that make a tale worth telling (and retelling): an historical basis, intrigue and murder, a quest for justice, and a spine-tingling resolution. Not surprisingly, the "ghost" seems to have taken on a life of its own.

REFERENCES

Brunvand, Jan Harold. 1978. *The Study of American Folklore: An Introduction.* 2d ed. New York: W. W. Norton.

Cranfield, Louis. 1963. Was Australia's greatest ghost story a hoax? *Chronicle* (Adelaide, S.A.), 24 October.

Cusack, Frank, ed. 1967. *Australian Ghosts.* Sydney: Angus & Robertson.

Davis, Richard. 1998. *The Ghost Guide to Australia.* Sydney: Bantam.

Finucane, R. C. 1984. *Appearances of the Dead: A Cultural History of Ghosts.* Amherst, N.Y.: Prometheus Books.

Fisher's ghost. 1853. *Household Words* 7: 6–9.

Fisher's Ghost appears but crowd disappoints. 1957. *Campbelltown-Ingleburn News* (Campbelltown, N.S.W.), 18 June.

Fowler, Verlie. 1991. *Colonial Days in Campbelltown: The Legend of Fisher's Ghost.* Rev. ed. Campbelltown, N.S.W., Australia: Campbelltown & Airds Historical Society.

Ghost that trapped a murderer? 1955. *Sun-Herald* (Sydney, N.S.W.), 3 July.

Illusionist brought Fisher's ghost to Pitt St. playhouse. 1984. *Daily Mirror*, 22 March [clipping in Campbelltown City Library's vertical file].

Lang, John. N.d. [1859?]. *Botany Bay or True Stories of the Early Days of Austra-*

lia. Excerpted in *Australian Ghosts,* edited by Frank Cusack, 1–24. Sydney: Angus & Robertson.
Lee, C. N. 1963. Another ghost version. *Campbelltown-Ingleburn News* (Campbelltown, N.S.W.), 12 February.
Nickell, Joe. 2001. Mysterious Australia. *Skeptical Inquirer* 25, no. 2 (March/April): 15–18.
Supposed murder. 1826. Notice in *The Australian*, 23 September [cited in *The Ghost Guide to Australia,* by Richard Davis, 16–18. Sydney: Bantam].
Supreme Criminal Court. 1827. Proceedings published in *Gazette* (Sydney), 5 February.

37

Ghostly Portents in Moscow

When I knew I would be visiting Russia in 2001, I began to wonder what paranormal mysteries there might be to investigate. Not surprisingly, according to some sources Russia is a haunted place, and ghostly phenomena seem similar to those reported elsewhere.

For example, a photograph of a "Moscow ghost" illustrates the section on Russia in *The International Directory of Haunted Places* (Hauck 2000, 129–31). However, the anomalous white shape at the bottom of the nighttime tourist photo seems consistent with other photographs in which foreign objects in front of the lens have bounced back the flash to create various ghostly effects: strands of "ectoplasm" (caused by a wandering wrist strap), "orbs" (due to water droplets, dust particles, etc.), or, in the case of the "Moscow ghost," apparently a careless fingertip!

Then there are the apparitions. Hauck (2000) states that at the Kremlin (Figure 37-1), the specter of Ivan the Terrible "is seen" (using

FIGURE 37-1. **The Kremlin, Moscow's ancient fortress, is allegedly a haunted place.**

the present tense); another source ("Ghosts" 1997) states that images of the fearsome czar "appeared many times" (employing the past tense). Both sources agree that the most appearances occurred in 1894, before Nicholas II's wedding to Alexandra, when he was to accede to the throne following the death of his father. Supposedly Ivan's phantom "brandished his heavy baton as sinister flames danced about his face."

The Kremlin is also allegedly haunted by the ghost of Lenin (1870–1924), the Russian revolutionary. In 1961 his spirit supposedly contacted a medium to tell her he despised sharing his mausoleum with dictator Joseph Stalin (1879–1953), "who caused the Party so much harm." The medium, a Communist heroine devoted to Lenin and imprisoned by Stalin, revealed Lenin's alleged wishes in an address to the Party Congress, and the next night Stalin's body was removed and reinterred elsewhere. Also, during the power struggle of 1993, Lenin's ghost was reportedly seen pacing anxiously in his former office and other government sites (Hauck 2000, "Ghosts" 1997).

What are we to make of such tales? First of all, the apparitions seem typical of those reported elsewhere, and may have the same explanations: that is, they may be due to a welling up of the subconscious during daydreams or other altered states of consciousness, and may be triggered in part by the power of suggestion, especially among certain highly imaginative persons (Nickell 2000).

Moreover, the reported encounters show an obvious tendency for Russian ghosts to serve as omens. People perceive ghosts in terms of their own cultural attitudes, which vary in different places and times. According to R. C. Finucane's *Appearances of the Dead: A Cultural History of Ghosts* (1984):

> Each epoch has perceived its specters according to specific sets of expectations; as these change so too do the specters. From this point of view it is clear that the suffering souls of purgatory in the days of Aquinas, the shade of a murdered mistress in Charles II's era, and the silent grey ladies of Victoria's reign represent not beings of that other world, but of this.

As purposeful as Shakespeare's ghosts (recall how the specters of Hamlet's father and Banquo were driven by revenge), those of Russian historical figures tend to function as portents in anxious times. I toured the Kremlin and stood outside Lenin's tomb in Red Square; I saw no ghosts, but I nevertheless could not escape feeling the impress of the often-troubled past of a great country. Such personages as Ivan the Terrible and Lenin, magnified by history and legend, loomed large—at least in my own haunted thoughts.

REFERENCES

Finucane, R. C. 1984. *Appearances of the Dead: A Cultural History of Ghosts.* Buffalo, N.Y.: Prometheus Books.

"Ghosts." 1997. *The St. Petersburg Times*, November 3. Retrieved 1 October 2001 from www.sptimesrussia.com

Hauck, Dennis William. 2000. *The International Directory of Haunted Places.* New York: Penguin.

Nickell, Joe. 2000. Haunted inns. *Skeptical Inquirer* 24, no. 5 (September/October): 17–21.

38

Mystique of the Octagon Houses

Phrenology, psychical claims, quack medicine—these and other fringe ideas have interesting connections to the octagon-house fad of the nineteenth century. I have visited a few of these historic eight-sided buildings, including a "haunted" one at Genesee Country Village in Mumford, New York. Individually and collectively, the structures present many features of interest.

Nearly Circular

Octagonal structures have a long history. During the Middle Ages they were often attached to churches to enclose baptistries. In America, from 1680 to 1750, octagonal churches were built in Dutch settlements along the Hudson River. Virginians George Washington and Thomas Jefferson were also attracted to the distinctive shape: Washington had eight-sided garden houses constructed at Mount Vernon, and Jefferson erected an octagonal retreat, Popular Forest (built 1806–1809), in Lynchburg ("Fowler's" 1992; Schmidt and Parr n.d., 7, 109).

The octagon plan had the primary advantage of a round structure—that of enclosing more interior space, for its circumference, than other shapes—but was easier to construct. Its obtuse angles eliminated dark corners (an idea that reportedly appealed to Jefferson, who included an octagonal guest room at Monticello [Shribman 2000]). The octagon also made possible a compact floor plan, with rooms easily accessible by a central hallway. Topped with a glass cupola, the octagon's core had natural lighting and was ideal for a winding stair.

The octagon's most ardent proponent, Orson Squire Fowler (1809–1887), also believed that the shape was a more healthful form. At a time when most people closed their windows to keep out "night vapors," Fowler promoted fresh air and argued that octagons provided better overall ventilation. Also, the larger rooms that his plans promoted allowed the air to circulate naturally, so that one was not faced with a choice of either repeatedly breathing the same air or sleeping in the draft of an open window (Fowler 1853; Kammen 1996).

"Fowler's Folly"

One writer described Fowler as a "compound of fanaticism, superficiality, and quackery"; another commented, "Orson Fowler was all of those things, but he was also ahead of his time" (Kammen 1996). With his younger brother Lorenzo, Fowler is best known for promoting phrenology, the pseudoscience of reading character from the contours of the skull. The brothers wrote numerous books on the subject and allied topics, and also published the *American Phrenological Journal* and later the *Water Cure Journal*. (The latter, briefly edited by Walt Whitman, promoted hydropathy, a treatment that supposedly cured all diseases by the internal or external use of large amounts of water.)

Fowler sensibly opposed the tight lacing of ladies' corsets; advocated exercise and good diet along with fresh air to promote vitality;

and fervently endorsed all forms of what he considered self-improvement. However, he also dabbled in "magnetism" (i.e., mesmerism) and published tracts extolling not only phrenology and medical quackery, but also clairvoyance and spiritualism. He fancied himself an expert on marriage and sex education, based on phrenological principles as well as his own experience from three marriages. He was also a temperance advocate, opposing all stimulants (including tea). Curiously, though, unlike many of those who shared his reformist views, he never joined the abolition movement (Kammen 1996; Fowler and Fowler 1855).

Octagons appealed to freethinkers and individualists like Fowler and can be seen as an expression of the "ultraist" and "perfectionist" sentiments that characterized western New York during the nineteenth century. Termed the "burned-over district," the area was "[a] hotbed of revivalism, millennialism, and perfectionism." Indeed, the same region that gave rise to the Shakers, Oneida Community (a free-love sect), Millerites (an end-of-the-world cult), and Mormons—as well as the Spiritualists and others—"also saw the octagon as a perfect and natural form of architecture" (Genesee n.d.).

In 1848, Fowler published the first version of his octagon-promoting book *A Home for All* and began construction of his own such home overlooking the Hudson River near Fishkill, New York. It was still unfinished five years later, but Fowler moved his family into the four-story, sixty-room house. Topped with an octagonal glass-domed cupola, the grout-walled house was an impressive sight, but neighbors soon dubbed it "Fowler's Folly." Fowler's finances suffered from the Panic of 1857, and he was forced to move out and rent the dwelling to a realtor. The latter turned it into a boarding house, but the walls leaked and a contaminated well caused a typhoid outbreak. The house subsequently passed through a succession of owners until, in 1897, it was condemned and demolished by dynamite (Kammen 1996, "Fowler's" 1992).

In the meantime, Fowler had inspired the construction of hundreds of octagon homes, and a group of utopian-minded leaders even

FIGURE 38-1. The Hyde octagon house at Genesee Country Village, Mumford, N.Y.

FIGURE 38-2. Joe Nickell looking for ghosts in the Hyde house's cupola at the head of the winding stairs. (Photograph taken for the author by Benjamin Radford.)

proposed an "Octagon City" in the Kansas Territory (this city failed to materialize). The 1857 Panic, which resulted from overspeculation in real estate and railroads, caused a decline in the octagon fad, which further declined with the onset of the Civil War. Nevertheless, octagons continued to be built until about the end of the century, especially in New York ("Fowler's" 1992).

The Hydes' House

One of these octagon homes was built at Friendship, New York, circa 1870, for Erastus C. and Julia E. (Watson) Hyde (Figures 38-1 and 38-2). It now stands as part of Genesee Country Village at Mumford, New York. The village is a 200-acre site occupied by historic buildings that have been transported from various locales and set as a recreated hamlet, complete with outlying vintage farm buildings. It is a wonderful accomplishment.

Following the Civil War, in which then-Corporal Hyde had been a drummer in an infantry company, he returned to Friendship to resume labor on his father's 100-acre farm. In 1869 he obtained an interest in a steam-powered shingle mill, and by 1870 he had also acquired a wife and an octagon house. The Hydes were both musicians and also shared interests in medicine and religion. In 1881 they went to Philadelphia, where Erastus enrolled in the Hahnemann Medical College, which taught homeopathy. (Homeopathy is based on the notion that "like cures like," and holds that extremely dilute doses of the substances that produce certain symptoms can alleviate those symptoms. The more diluted the dosages the better, homeopaths believe, with the result "that not a single molecule of the original substance remains" in a homeopathic preparation [Randi 1995, 159–60].)

While Hyde studied medicine, his wife taught music at a Philadelphia conservatory. After he received his degree, in 1884, she studied Methodism and became an ordained minister of that faith (Genesee n.d.). The couple returned to the octagon home at Friendship, and in one of the triangular rooms on the first floor Dr. Hyde established his medical office. (Although today a phrenological bust is exhibited in the parlor, it was presumably placed there because of the octagon's connection to Orson Fowler, and not to suggest that Hyde also was a phrenologist.)

Spirit Communication

Ardent Spiritualists, the Hydes also had a home (and Dr. Hyde another medical office) at Lily Dale, the Spiritualist village. Hyde served on the board of directors from 1895 to 1901, and was on the medical staff of the nearby Lily Dale Sanitarium as a "Physician and Electro-therapist" (as reported in the Spiritualist *Banner of Light*, 25 July 1895). Mrs. Hyde wrote poetry, taught music, held classes in her home on spiritual self-development, and helped organize both a temperance society and a naturalist club for children (LaJudice 2001). (The Hydes were childless, but reportedly adopted several children and aided others with scholarships [Genesee n.d.].)

In the mid-1890s, there was a scandal at Lily Dale involving phony spirit manifestations by one Hugh Moore. (Moore is described in accounts of activities at Lily Dale as a "trumpet medium" in 1894 and "the materializing medium" in 1895 [LaJudice 2001].) The scandal is evidenced by a letter from Julia Hyde (1896) to a friend. She wrote, in part:

> I do not anticipate that the cloud shadowing Lily Dale will be permanent. I hope it will result in a wholesome weeding out of fraud and "ringing in of the true." I do not question the mediumship of many of these people, but their greed of gain prompts them to descend to actions that true spiritualists can not condone nor sanction; but that does not do away with the *real* in our cause. I do not, and never did believe that physical manifestations are the basis of spiritual belief. They are diametrically opposed to each other in expression . . . .

Later she said:

> Yes, I do not question but that Hugh Moore was a medium; no doubt they all are more or less, but when they can not get enough *genuine* to satisfy the morbid appetites of marvel seekers, they tack on the false and then cry continually about "conditions," to keep away honest investigation, and cover up their tracks . . . . Hugh Moore is said to be keeping a whiskey saloon in Dayton, O. He jumped his bail in Cincinnati, and left his confederates, who helped "play spirits," unpaid . . . .

Nevertheless, Mrs. Hyde asserts that she is "a firmer spiritualist today than ever before, because of my hope that false conditions are passing out, and being overcome, and the genuine will triumph at last."

Although "it was said" that Julia Hyde conducted séances in the octagon house parlor (Bolger 1993, 98–99), my research at Lily Dale turned up no evidence that she was herself a medium, and the previously quoted letter written by her does not suggest so. Of course she could have hosted séances conducted by mediums in her octagon home, but the available evidence suggests that her spiritualist involvement was largely limited to Lily Dale. Nevertheless, the supposed séances, coupled with her dying within two days of Dr. Hyde's death, led to "the belief even among sensible people" that "their departed spirits frequented the old, oddly-shaped house" (Bolger 1993). Its reputation as haunted really arose only after the house at Friendship was abandoned; abandoned houses seem always to invite notions of ghosts and hauntings. Today the administrators of Genesee Country Village commendably adopt the modern view that promoting claims of haunting at a historic site is unprofessional. Three tour guides I spoke with were skeptical of reports of ghosts in the octagon house, as none had had any haunting experiences. One insisted that she *would* not, since she was a good Catholic who did not believe in such things; another docent, who had never seen ghostly phenomena at any of the four or so reputedly haunted village buildings, jokingly added that if ghosts actually existed they should appear to her so as to end her skepticism (Nickell 2000; 2001).

Other Octagons

At Lily Dale is another octagon building, originally topped with a cupola. It was constructed in 1890, so it is tempting to think that the design was suggested by the Hydes, who were prominent residents there as early as 1888. Described in one article as "an octagonal séance house" (Kammen 1996), it was used for séances in the 1940s but was originally used for classes in social etiquette, dancing, and the like (LaJudice 2001). It now houses the Lily Dale Mediums' League.

Another octagon I have visited in the "burned-over district" is the Rich-Twinn Octagon House at Akron, New York, which is on the National Register of Historic Places. Built in 1849–1850, just after Orson Fowler published the first edition of *A Home for All* in 1848, it may well have been among the first fruits of Fowler's octagon advocacy.

So distinctive are the octagons that their popularity has resulted in at least one misnomer: The historic (and supposedly haunted) four-story architectural landmark in Washington, D.C., known as "The Octagon" actually has only six sides. Built between 1797 and the early 1800s, it was given its unusual shape to fit an acute angle at the juncture of two thoroughfares (Alexander 1998, 98–104).

Although many of Orson Fowler's ideas were discredited even before his death in 1887, his octagon concept was subsequently expanded upon by Buckminster Fuller (1895–1983) in creating his celebrated geodesic dome. (Once, in 1967, with a handful of other college students, I sat literally at "Bucky's" feet to hear and discuss his fascinating futuristic ideas.) Like Fowler, Fuller was searching for a design for a healthy and inexpensive home, and his geodesic house was intended for mass production at a retail cost of 50 cents per pound (Kammen 1996)!

Fowler's greatest legacy remains in the form of numerous surviving octagons. (A state-by-state list is provided in Schmidt and Parr [n.d.].) More exist in New York State than anywhere else. Many of them, including examples at Lily Dale, Akron, and Genesee Country Village, have been lovingly restored, thus standing as "monuments to their builders' desire to create a special and healthy home—and to the ideas of Orson Squire Fowler" (Kammen 1996).

REFERENCES

Alexander, John. 1998. *Ghosts: Washington Revisited.* Atglen, Pa.: Schiffer Publishing.

Bolger, Stuart. 1993. *Genesee Country Village: Scenes of Town & Country in the Nineteenth Century.* 2d ed. Mumford, N.Y.: Genesee Country Village.

Fowler, Orson S. 1853. *A Home for All, or the Gravel Wall and Octagon Mode of Building.* Reprinted 1973 as *The Octagon House: A Home for All.* New York: Dover.

Fowler, O. S., and L. N. Fowler. 1855. *The Illustrated Self-Instructor in Phrenology and Physiology* . . . . New York: Fowlers and Wells.

"Fowler's follies." 1992. *Colonial Homes*, August, 68–71.

Genesee Country Village. N.d. Unpublished typescript on the Friendship Octagon House. Mumford, N.Y.: The Genesee Country Museum.

Hyde, Julia. 1896. Letter to a Mrs. Peck, written at Lily Dale, 25 April [typescript text in Genesee n.d.].

Kammen, Carol. 1996. Orson Squire Fowler and the octagon. *Heritage* 12, no. 2 (Winter): 4–15.

LaJudice, Joyce. 2001. Personal communication and *Lily Dale Chronicle* (computer database in progress, compiling information on Lily Dale from historical sources such as the *Banner of Light*) by the Lily Dale historian.

Nickell, Joe. 2000. Case-file notes, Hyde Octagon House, 2 July.

———. 2001. Case-file notes, Genesee Country Village, 3 August.

Randi, James. 1995. *The Supernatural A–Z: The Truth and the Lies.* Great Britain: Brockhampton Press.

Schmidt, Carl F., and Philip Parr. N.d. [1978?]. *More About Octagons.* N.p.: Privately printed.

Shribman, David. 2000. Slavery casts shadow on U.S. history. *Buffalo* (N.Y.) *News*, 6 July.

39

Weeping Icons

A paranormal phenomenon enjoying favor in the new glasnost of Russia is that of "miraculous" icons—notably one that was reported to be weeping in a Moscow church in 1998.

The Russian Orthodox Church has a tradition of venerating icons (from the Greek *eikon*, "image"), which are painted on varnished wood panels and over time acquire a dark patina from candle smoke. Russian icons were produced in greatest number at Kiev, where Christianity took root in 988 (Richardson 1998, 222). Perhaps because they naturally depicted holy subjects and miraculous events—such as the imprinting of Jesus' face on Veronica's veil, shown in a fourteenth-century icon that I viewed in the Tretyakov Gallery—they seemingly began to work miracles themselves.

The claim that an effigy is in some way *animated* (from *anima*, "breath") crosses a theological line from *veneration* (reverence toward an image) to *idolatry* (or image worship) in which the image itself is regarded as the "tenement or vehicle of the god and fraught with divine influence" (*Encyclopedia Britannica* 1960). Nevertheless, reports of weeping, bleeding, and otherwise animated figures continue. In one modern case, in Sardinia, in which a small statue wept blood, samples were analyzed; the DNA proved to be that of the statue's owner. Yet her attorney reasoned, "Well, the Virgin Mary had to get that blood from somewhere" (Nickell 1997).

"Salty tears" were said to flow from another image in Pavia, Italy, in 1980. No one witnessed the initial weeping, only the flows in progress, and the owner seemed to be alone with the figure (a small plaster bas-relief) whenever it wept. Soon, suspicious persons, peeking through the windows and a hidden hole in an adjacent apartment, saw the owner apply water to the bas-relief with a water pistol (Nickell 1997)!

In 1996, in Toronto, pilgrims were charged $2.50 at a Greek Orthodox church to view an icon that "wept" oil. As it happened, the priest had once presided over another "weeping" icon in New York, and had even been defrocked for working in a brothel in Athens. I was involved in the case twice, the second time at the request of the parent church. With a fraud-squad detective standing by, I took samples of the oily "tears" for the Center of Forensic Sciences. The substance proved to be a nondrying oil, as expected; its use is an effective trick, since one application remains fresh-looking indefinitely. Because no one could prove who perpetrated the deception the case fizzled, but the church's North American head pronounced it a hoax (Nickell 1997, Hendry 1997).

One interesting feature of the exuding icons is the variety of substances involved (blood, salt water, oil, etc.), as well as the different effects (e.g., weeping tears, sweating blood, exuding oil). When the cases are collected and compared, some trends become apparent. In Catholicism, the images tended to yield blood or watery tears until relatively recently, when—more in line with the Greek Orthodox tradition (possibly due to a number of oil-weeping or -exuding icons at such churches that received media attention)—there has been a shift to oil (see, e.g., Nickell 1999).

For instance, among the reputed miracles that attended a comatose girl at a Catholic family's home in Massachusetts in the 1990s were oil-dripping statues and images. Analysis of one sample of oil found that it was 80 percent vegetable oil and 20 percent chicken fat, according to *The Washington Post*, which ordered the test. Such a concoction would have been readily available in a home kitchen (Nickell 1999).

Interestingly, icons in the Russian Orthodox tradition seem, rather uniquely, to exude myrrh—or rather, apparently, myrrh-scented oil. *Myrrh* is a fragrant gum resin used in making incense, perfume, and herbal medicines, and in ancient times it was also employed in embalming. (For instance, it was one of the spices used in Jesus' burial, interspersed with his linen wrappings [John 19:39–40].) Indeed, in St. Petersburg in 1998, when an unidentified mummy began to exude a myrrh-like substance, it was regarded as a miracle that helped identify the remains as the lost relics of a sixteenth-century saint, Alexander of Svira. His relics had disappeared in 1919 when Bolsheviks seized them during repressive actions against the church. "According to Orthodox tradition," explains one source, "the appearance of fragrant liquids on relics is a miracle and means they belong to a saint" (Laguado 1998). Although forensic experts cautioned against a rush to judgment, priests were satisfied that droplets of the substance between the mummified toes were myrrh and therefore evidence of a miracle. They seem to have ignored the possibility that myrrh could simply have been used in the embalming.

Given this cultural backdrop, it is not surprising to find that Russian Orthodox icons—when they are in a reputedly miraculous mode—tend to yield myrrh as the substance of choice. This is true even of icons at Russian Orthodox monasteries in the United States. In 1985, an icon in Blanco, Texas, was discovered "weeping Myrrh." The Christ of the Hills Monastery subsequently produced a brochure advertising itself as a "Shrine of the Blessed Virgin Mary," claiming "She weeps tears for all mankind." Anointment with the tears from this icon had produced "great miracles," including "cures of cancer, leukemia, blindness, mental illness" and so on ("Shrine" n.d.).

Similarly, in 1991, an icon that now reposes in a Russian Orthodox monastery in Resaca, Georgia, commenced to "exude myrrh." It welled in the eyes of the Virgin Mary and was held to be "the external tears of the Mother of God, revealed in the Weeping Ikon"—according to an advertising brochure circulated by the monastery ("All-Holy" n.d.).

In 1998, in Moscow, an icon portraying the last czar, Nicholas II, reportedly produced myrrh almost daily after a parishioner brought it to the church on 7 November, the date of the Russian revolution in 1917. Nicholas—along with the czarina, their children, servants, and a personal physician—was assassinated on the night of 16 June 1918. (Eventually their remains were discovered, identified through DNA, and given a funeral in 1998.) ("Church" 1999).

When I learned I was going to Moscow, I resolved to try to track down the lachrymose icon of Czar Nicholas. Subsequently, friend and colleague Valeriĭ Kuvakin and I made our way by bus and Moscow's excellent subway system to one of the oldest districts in the city, where we soon found the onion-domed church called the Church of Nikola in Pyzhakh. There, as we looked around the interior, we observed the usual proliferation of icons, displayed on the iconastasis (a high screen that separates the sanctuary from the nave) and elsewhere. At least one depicted a weeping female saint, and I wondered if such depictions might have sparked the idea of "actual" weeping icons. On making inquiry about taking photographs, we learned that they were prohibited, although a few rubles later we had permission to take a single picture. We also obtained a devotional card featuring the icon of the czar (Figure 39-1).

We were surprised to learn, according to the text on the reverse of the card, that the miraculous icon was only a color photocopy. The

FIGURE 39-1. **Previously myrrh-exuding icon of the last czar, Nicholas II, shown on a devotional card.**

original was painted by an American artist commissioned to glorify "the suffering czar." In 1987 a monk brought it to Russia, where photocopies were made, and one of those photocopies was received in Moscow in 1998. After prayers were made on the czar's behalf, the picture became fragrant on 6 September and began weeping on 7 November. Actually. the word used translates as "myrrhing"—that is, "yielding myrrh." The picture went on tour in Russia, Belorussia, and Serbia, and more than a dozen "healing miracles" were attributed to "the myrrhing image of our last czar," and thousands of believers who prayed to him supposedly received help and support.

Unfortunately, when we visited the church the icon was no longer weeping. Nevertheless, people were coming into the sanctuary every few minutes to view the icon: typically they kissed the glass that covered it and prayed, though a few even prostrated themselves before it. When I was able to get a look at the icon myself, I could see that, indeed, it was merely a cheap facsimile. I sought to learn more about the circumstances of the previous "myrrhing," but Valeriĭ's questions to the church staff were met with obvious suspicion (because, Valeriĭ concluded, we were not showing devotion). We therefore learned little apart from press reports and the text of the devotional card.

The staff's reaction made me suspicious in turn, as I have more than once found a wary attitude masking pious fraud. Further suspicions are raised by the fact that, as we have seen, other "weeping" icons have been proven or suspected to be fakes; that Russian Orthodox icons exhibit a culturally distinct form of the "miracle" ("myrrhing"); and that the phenomenon occurred at a time when there was a campaign to bestow sainthood on Czar Nicholas II and his family. The patriarch of the church, Alexy II, opposed the canonization, stating that the imperial family were undeserving because of their poor leadership of both church and state ("Church" 1999). The "miracle" seems an attempt to counter that view by faking a semblance of divine approval.

REFERENCES

"The All-Holy Theotokos." N.d. Brochure of the [Russian] Orthodox Monastery of the Glorious Ascension, Resaca, Georgia.

Church to test Moscow icon. 1999. AOL News (AP), 30 January.

Encyclopædia Britannica. 1960. s.v. "Idolatry."

Hendry, Luke. 1997. "Weeping" icon called a fake. *Toronto Star*, 28 August.

Laguado, Alice. 1998. Orthodox Church sanctifies mummy. *Arizona Republic*, 22 August.

Nickell, Joe. 1997. Those tearful icons. *Free Inquiry* 17, no. 2 (Spring): 5, 7, 61.

———. 1999. Miracles or deception? The pathetic case of Audrey Santo. *Skeptical Inquirer* 23, no. 5 (September/October): 16–18.

Richardson, Dan. 1998. *Moscow: The Rough Guide.* London: Rough Guides, Ltd.

"Shrine of the Blessed Virgin Mary." N.d. Brochure of the Christ of the Hills Monastery, Blanco, Texas.

40

Spiritualist's Grave

Among the sites that supposedly make Australia "a very haunted continent" is the Rookwood Cemetery in Sydney (*International* 2000). One of the graves there has a profound link to spiritualism and once attracted famed magician Harry Houdini. It is the burial place of William Davenport (1841–1877), one of the notorious Davenport Brothers and the subject of an interesting story.

Ira and William Davenport debuted as spiritualists in Buffalo, New York, in 1854, when they were yet schoolboys (aged 15 and 13 respectively). Soon they were touring the world giving demonstrations of alleged spirit phenomena. While the pair were securely tied in a special "spirit cabinet," the "spirits" played musical instruments and performed other "manifestations" in darkened theaters.

On July 1, 1877, while they were on tour in Australia, the long-ailing younger brother William died and was buried at Rookwood. Decades later, in 1910, while Houdini was himself on tour there (and incidentally entered Australian history by becoming the country's first successful aviator), the great magician/escape artist paid a visit to the grave, accompanied by magicians Allan Shaw and Charles J. Carter (Christopher 1976, 60–83).

Houdini (1924, 17–37) found the grave "sadly neglected" and so, he wrote, "I had it put in order, fresh flowers planted on it and the stone work repaired." Subsequently, when Houdini met the surviving Davenport brother, Ira was so moved by Houdini's act of kindness that he confessed the brothers' tricks, even teaching his fellow escapologist "the famous Davenport rope-tie, the secret of which," Houdini noted, "had been so well kept that not even his sons knew it."

My own interest in the Davenport brothers was renewed when I was able to help bring to light the contents of their personal scrapbook (Nickell 1999). I had continued my interest in the duo by locating and visiting Ira's grave in Mayville, New York. Now, finding myself in Sydney, I determined to recreate Houdini's visit to William's grave. I was accompanied by Peter Rodgers and by another magician, Kent Blackmore (both of whom had visited the site in 1983).

The Rookwood Cemetery is huge, and thus it took us some time to relocate the grave (in the Church of England Necropolis, section E, grave number 848). Armed with weed clippers and a bouquet of fresh flowers, we soon made the site presentable once again. The gravestone's inscription reads: "Sacred to the dearly beloved memory of William Henry Harrison Davenport of the Davenport Brothers. Born at Buffalo U.S.A. Feb. 1st 1841 and who departed this life July 1st 1877 after a long and painful illness which he bore with great courage and gentleness. May he rest in peace. Erected by his loving wife." On the reverse of the stone is inscribed: "To William, / from his brother Ira. / Dear brother I would learn from thee / And hasten to partake thy bliss. / To thy world, Oh welcome me / As first I welcomed thee to this."

Like the trio who preceded us in 1910, we three magi posed for

FIGURE 40-1. **Trio of magicians—Joe Nickell, Peter Rodgers, and Kent Blackmore—recreating the 1910 gathering of Houdini and friends at the grave of spiritualist William Davenport.**

photographs to record the event (Figure 40-1). Alas, neither William Davenport's nor any other spirit put in an appearance, as far as we could tell. Nevertheless, it was an occasion to recall those who lived in earlier times and to reflect on how things have since changed and yet remained much the same. For instance, although the physical manifestations of spiritualism's earlier era have largely been supplanted by mental mediumship (as practiced by spiritualists like John Edward and James Van Praagh [see chapter 21]), the attraction to alleged spirit communication continues.

So does the interest in other paranormal claims. Although during my time Down Under I pursued several mysteries that had a decidedly Australian flavor, they nevertheless represented many of the same themes—hauntings, monsters, etc.—that are found virtually everywhere. How familiar is the strange, we might say, and even, considering Australia's distinctive offerings, how strange the familiar.

REFERENCES

Christopher, Milbourne. [1976] 1998. *Houdini: A Pictorial Biography.* Reprinted New York: Gramercy Books.

Houdini, Harry. [1924] 1972. *A Magician Among the Spirits.* Reprinted New York: Arno Press.

International Haunted Places. 2000. Retrieved 4 August 2000 from http://freehosting2.at.webjump.com/269be35db/ha/haunted-places/International.htm

Nickell, Joe. 1999. The Davenport Brothers. *Skeptical Inquirer* 23, no. 4 (July/August): 14–17.

41

Incredible Stories

Charles Fort and His Followers

Mystery-mongering sells. Why else would Barnes and Noble issue a 1998 edition of *The World's Most Incredible Stories: The Best of Fortean Times*? Originally published in London (Sisman 1992), this collection of oddities, anomalies, and occult claims is (as its subtitle indicates) in the tradition of Charles Fort. Fort (1874–1932) loved to challenge "orthodox" scientists with things they supposedly could not explain, like rains of fish or frogs (Fort 1941).

In the introduction to *Incredible Stories*, Lyall Watson paints a typically fortean, typically disparaging view of science: an endeavor that "claims to be objective" but is "inherently conservative and resistant to change," even a fundamentally "political process" that "depends on personal preference, upon the votes of a scientific jury—every member of which would be disqualified from any normal inquiry on the basis of blatant conflict of interest."

FIGURE 41-1. **Grave of Charles Fort (1874–1932). (Photo by Joe Nickell.)**

To Watson, what is needed is "a truly impartial investigator—a sort of scientific ombudsman—to provide the voice of reason, to speak out for curious individuals against the vested interests of those in authority." Fort fits the bill, says Watson, who seems to speak for forteans everywhere when he states, "I know that there is a vast field of unusual experience from all over the world, just waiting to be examined. The problem is that reports of it are, by their very nature, anecdotal, and therefore dismissed as unacceptable to science."

In fact, however, Charles Fort did not actually investigate reported occurrences. Having come into an inheritance that permitted him to sit comfortably and indulge his hobby, he spent his last 26 years scouring old periodicals for reports of mysterious occurrences, giving the distinct impression that he believed whatever was asserted was true—or at least suitable for taunting members of the scientific "priestcraft." Thus Fort is the poster boy for the limits of anecdotal evidence, and his armchair mystery-mongering attitude is continued by Adam Sisman, who selected and edited the 1992 collection of tales.

Some items in the collection are non-mysteries, like a runaway wallaby; and others are merely Ripleyesque: a six-legged lamb (Sisman 1992, 10), identical twins who gave birth on the same day (20), and a stationers' shop named Reid & Wright (22). There is even some genuine skepticism, such as with the reports of "wolf children" who are acknowledged as probably being mentally and physically handicapped (102). A token handful of hoaxes and urban legends—that is, ones actually recognized as such—are also included (188–91).

Nevertheless, numerous mysteries touted in the book as allegedly paranormal are bogus. For example, the front cover portrays "Psychic Katie," with copper foil apparently having materialized on her body. Yet this, along with one of Katie's other feats—producing glass gems from her eye—was easily duplicated by CSICOP investigators for an October 1990 episode of the *Unsolved Mysteries* television program. Slow-motion study of the gem feat showed that the object was apparently hidden between her fingers (Nickell 1997).

Regarding Atlanta's 1987 "House of Blood" mystery, the fortean book relates how the elderly residents had described blood springing from the floor "like a sprinkler" and claims that it had appeared "in narrow spaces virtually impossible for a person to reach" (Sisman 1992,

93). In fact, a *Skeptical Inquirer* article from the spring of 1989 quoted a police detective as suggesting that the affair was a hoax, and a subsequent investigation I conducted in 1994 provided corroborative evidence. Expert blood-pattern analysis of police photographs revealed that the blood had been squirted *onto* surfaces, rather than having spurted *from* them as the residents had asserted (Nickell 1995, 92–97).

In describing "The Aerial Fakir," the Sisman book boasts that Subbayah Pullavar once levitated and "remained horizontal in the air for about four minutes." An accompanying photograph documents the feat (166–67). Unfortunately, in stage magician's parlance, the effect was not a *levitation* but only a *suspension*, since one hand rested on a rod wrapped in cloth. The secret of such suspensions, consistent with the details of Pullavar's performance, is illustrated in conjuring texts (e.g., Gibson 1967, 81–83).

In a section called "Psychic Powers," readers are told that "after cutlery-bending Yuri [sic] Geller's first British broadcast on 23 November 1973, children began to discover their own paranormal powers." They could bend metal objects like nails and keys, and perform other feats, "seemingly just by thinking hard" (170). A photograph in physicist John Taylor's 1975 book *Superminds* showed a seven-year-old boy's supposed psychokinetically bent fork and spoon. Alas, as reported by Martin Gardner (1979–1980), the little psychokinetic marvels were actually exhibiting "kindergarten principles of deception." Observed secretly, the children simply bent the metal in the usual way. A boy used both hands to bend a spoon, while a little girl placed the end of a rod under her foot.

Under "Poltergeists," the forteans include the 1984 case of Tina Resch, the Columbus, Ohio, 14-year-old who seemed startled by airborne telephone receivers and other flying objects (Sisman 1992, 77). Although reporters and parapsychologists were duped, some photographs and television newstapes captured Tina in the act of toppling a lamp and producing other effects, and a television technician saw her surreptitiously move a table with her foot. Investigator James Randi characterized her at the time as a disturbed teenager (Randi 1985). A decade later, Tina Resch Boyer was sentenced to life imprisonment for the murder of her three-year-old daughter (Frazier 1995).

The mysteriously swirled patterns in southwest English grain fields known as *crop circles* are discussed from various perspectives, but the evidence for hoaxing gets short shrift: "Numerous attempts by crusading sceptics and newspapermen," states the book, "have failed abysmally to mimic the crop circle phenomenon, which is widely perceived as a hoax." The book concludes: "If it *is* some kind of practical joke, then the organization behind it outstrips the Mafia, KGB and Illuminati combined" (Sisman 1992, 110). Actually, in 1991, "two jovial con men in their sixties" admitted that they had launched and nurtured the hoax, in which they had been followed by many others in a bandwagon or copycat effect. The hoaxers quickly fooled circle "experts" who declared bogus patterns authentic (Nickell and Fischer 1992).

As these examples show, many of the claims in *Incredible Stories* do not withstand scrutiny. In addition, numerous tales lack specific names, places, dates, or source citations that might make further investigation possible. Such accounts suffer a lack of credibility as severe as the several pieces that rely on absurd tabloid sources like *Weekly World News*.

And then there are the simply outrageous assertions, like the unqualified statement that in 1951 Mary Reeser of St. Petersburg, Florida, "spontaneously combusted." Noting "evidence of the extraordinarily fierce heat, inexplicably contained," the book offers a dubious suggestion that there might have been a connection with "an intense geomagnetic storm" (Sisman 1992, 57). As the forteans could have learned from the Summer 1987 *Skeptical Inquirer*, Mrs. Reeser's death was not so mysterious. She was last seen smoking a cigarette after having taken sleeping pills—hers was thus an accident waiting to happen. The large stuffed chair she sat in and her own considerable body fat obviously contributed to the destruction, and the fact that the floor and walls of her efficiency apartment were made of concrete doubtless limited the fire's spread. By leaving out such details, *Incredible Stories* undermines fortean arguments against the scientific method.

The forteans must know that it takes little effort to launch an incredible claim, whereas serious, prolonged investigation is frequently required to get to the bottom of a mystery. Although there *are* legitimate enigmas that should not be dismissed out of hand, forteans have a responsibility not to make frivolous claims. Like the boy who cried wolf, they may find themselves without credibility.

REFERENCES

Fort, Charles. [1941] 1974. *The Complete Books of Charles Fort.* Reprinted New York: Dover.

Frazier, Kendrick. 1995. "Columbus poltergeist" Tina Resch imprisoned in daughter's murder. *Skeptical Inquirer* 19, no. 2 (March/April): 3.

Gardner, Martin. 1979–1980. The extraordinary metal bending of Professor Taylor. *Skeptical Inquirer* 4, no. 2 (Winter): 67–72.

Gibson, Walter. 1967. *Secrets of Magic Ancient and Modern.* New York: Grosset & Dunlap.

Nickell, Joe. 1995. *Entities: Angels, Spirits, Demons, and Other Alien Beings.* Amherst, N.Y.: Prometheus Books.

———. 1997. Mystery of the crystal tears. *Skeptical Inquirer* 21, no. 3 (May/June): 16–17.

Nickell, Joe, and John F. Fischer. 1992. The crop-circle phenomenon: An investigative report. *Skeptical Inquirer* 16, no. 2 (Winter): 136–49.

Randi, James. 1985. The Columbus poltergeist case. *Skeptical Inquirer* 9, no. 3 (Spring): 221–35.

Sisman, Adam. [1992] 1998. *Incredible Stories: The Best of Fortean Times.* Reprinted New York: Barnes & Noble.

Taylor, John. 1975. *Superminds.* New York: The Viking Press.

Index

abduction, by aliens. *See* UFO abductions
Ablaze! The Mysterious Fires of Spontaneous Human Combustion (Arnold), 10–11
Aborigines
Dreamtime, 291
yowie, 291–94
accelerator mass spectrometry (AMS) dating, Shroud of Turin, 196–97
acupressure, 215
acupuncture, 201
Adler, Alan, 194
Adolph and Rudolph, 86
African religion, and voodoo, 141
aftercatch, carnivals, 80
Afterlife, The (Randles and Hough), 237
Alexander of Svira, 326
Alfonso VI, king of Spain, 242
alien hybrid, 46–49
DNA testing, 49
aliens. *See* extraterrestrials
Allen, Steve, 253–54
Allen, Thomas B., 17
Alligator Boys and Girls, 84, 86
Allison, Dorothy, 253–56
discrediting, 254–56
Altea, Rosemary, 177, 286
altered states. *See* consciousness, altered states; fantasy-prone personality; hypnopompic/hypnagogic hallucinations; lucid dreaming; sleep paralysis
Alvarez, Luis, 230
American Institutes for Research (AIR), remote-viewing research, 64–65
American Museum (New York), 79
American Psychological Journal, 316
Amityville Horror: A True Story (Anson), 73, 74, 76
Amityville horror, 73–77
DeFeo murders, 73, 76–77
haunting, manifestations of, 73
as hoax, 75–77
Amityville II: The Possession (Anson), 73
anatomical wonders, carnivals, 85
Anderson, George, 177, 286

angels
  aliens as, 226
  psychological rationale, 171
animals
  carnival attraction, 88–89
  cryptids, 289–94
  mutilation and *El Chupacabras* (the goatsucker), 28–30
  pet mediums, 285–87
  pet psychics, 282–85
  psi research of, 281–82
  psychic pets, 279–81
  real versus swamp monsters, 166–74
  sacrifice, voodoo, 145, 146–47
  talking animals, 277–80
Anson, Jay, 74, 75
apparitions, 233–35
  and altered state of consciousness, 233–34
  See also ghosts; visitations, of dead
Appearances of the Dead: A Cultural History of Ghosts (Finucane), 313
apports, meaning of, 34
apports trick, 42–45
  method of trickery, 34, 43–44
Archer, Mike, 289, 291
Arnold, Larry E., 10–13
Association for Research and Enlightenment (ARE), 211, 213, 214
astral travel. *See* out-of-body experiences
astrology, 248–51
  challenging claims, 249–51
  signs of zodiac, 249
Atasha the Gorilla Girl, 78, 87–88
Atlantis, 104, 214
auditory hallucinations, and waking dreams, 219
aura
  of animals, 285
  scientific nature of, 201
Australia
  cryptids, 289–94
  Fisher's ghost, 304–9
  Hyde Park Barracks haunting, 271–76
  Rookwood Cemetery, 331–34
automatic writing
  and channeling, 66
  See also Writing, unknown sources
Aveni, Anthony F., 4
Aztecs, conversion to Catholicism, 41

Baima-Bollone, Pierluigi, 243–44
Baker, Robert A., 15, 136, 160, 171, 232, 264
Ballard, Chris, 182–83
bally, of carnival, 80
Banks, Elmore Lee, 158–59
Banner of Light, 320
Barber, Theodore X., 296–97
Bard, Ron, 252
Barnum effect, 283–84
Barnum, P. T., 79–80, 283–84
bearded ladies, 84
Bigfoot, size of tracks, 169
Bigfoot-like creatures
  sasquatch, 291
  tracks of, 168, 172–73, 293–94
  yeti, 291
  yowie, 291–94
billet reading, 38–40
  methods and trickery, 34, 39–40
birth defects
  carnival oddities, 83–85
  past explanations of, 48
birthmarks, 48
Bishop, Father Raymond J., 17, 23–24
Blackmore, Kent, 332–33
blade-box illusion, 88, 89
Blatty, William Peter, 17
Blavatsky, Helena P., as fantasy-prone personality, 297
Blitz, Antonio, 280
Blood and the Shroud, The (Wilson), 194
blowoff, carnivals, 91
BLT Research Team, 120

Index

bogeyman
origin of term, 171
voodoo, 145
Boggy Bayou monster, 173–74
Bogle, Michael, 273–74
Boliakiu, Kirill, 259
Bonnie and Clyde, 90
"booger" tales, 171–72
Book of Commandments (Smith), 302
Book of Mormon, The (Smith), 302
Booth, John Wilkes, mummy of, 90
Bowdern, Father William S., 17, 23
Bower, Doug, 118
Brodie, Fawn M., 300
Browne, Sylvia
cold readings, 286
errors, methods of covering, 183
as fantasy-prone personality, 137
Bucklin, Robert, 193
Budig, Ulrike, 49
bunyip, 289
burned-over district, octagon houses, 317, 322
bushranger, 275

Cabri, Jean Baptiste, 85
Cadena, Richard, 290
California, Blythe line drawings, 5
Camino: A Journey of the Spirit, The (MacLaine), 103
Camino pilgrimage, Santiago de Compostela, 103–5
Camp Chesterfield, 31–45
apports trick, 34, 42–45
billet reading scam, 34, 38–40
past exposés of, 32–37
photos of, 35, 36
spirit card writing trick, 40–42
undercover visit to, 35–44
camp meetings
fakery exposed, 32
for spiritualism, 32–33
See also Camp Chesterfield
Cardiff Giant, 90–91
card writing trick, 40–42
Carl, Christa, 284, 286–87
carnivals, 78–91
anatomical wonders, 85
animal acts, 88–89
Barnum sideshows, 79–80
blowoff, 91
compared to circus, 80
curios, 90–91
gaffed (faked) freaks, 86
human oddities, 79, 80–86
illusion show, 87–88
origin of, 78–79
outsiders (rubes), 91
tattooed persons, 85–86
ten-in-one shows, 80–83
torture acts, 87
wonder-workers, 87
carte de viste, Tom Thumb's wedding, 82
Carter, Charles J., 332
Cassadaga Lake Free Association, 32
Cassiopaeans, 49
Caterpillar Man, 84
Catholics and Catholic Church. *See* Roman Catholicism
Cayce, Edgar, 207, 211–14
biographical information, 211–12
readings, nature of, 212–14
cemeteries
necropolis (Spain), 107, 112
Rookwood Cemetery, 331–34
tomb of Marie Laveau, 152–56
Center for Inquiry-West, 129
Central Intelligence Agency (CIA), remote viewing, 61–71
cereologists. *See* crop circles
Chang and Eng, 85
channeling
and automatic writing, 66
See also clairvoyance; mediums
Charbonnet, Robbie, 167–68, 171

Chariots of the Gods?, 1–2
charismatic movement, speaking in tongues, 265–66
charms
  apports, 43–44
  gris-gris of voodoo, 143
  relics of saints, 102, 109
Charney, Geoffroy de, 188
Charney, Margaret de, 188–89
Chatworthy, Duane, 96
Chevalier, Ulysse, 198
chicken man, alien hybrid, 46–49
Childlight: How Children Reach Out to Their Parents from the Beyond (Theisen), 229
chimney effect, 163
chiropractic, forerunner of, 211
Chorley, Dave, 118
Chornyi, Kurií, 200
Christ of the Hills Monastery, weeping icon, 326–27
Christian Science, 134
Christopher, Kevin, 108, 119
Christopher, Milbourne, 19–20, 70, 250, 278
Church of Christ, 302
Church of Jesus Christ of Latter-day Saints, Joseph Smith, 296–303
Church of Nikola, weeping icon of, 327–29
cinematic neurosis, 17
Cipac, Marcos, 53
Circles Effect and Its Mysteries (Meaden), 120
circus, compared to carnival, 80
clairvoyance
  cold reading, 145
  forms of, 63
  of Marie Laveau, 145
  meaning of, 62
  medical intuitives, 207–16
  remote-viewing, 62–71
  See also mediums
Clement VII, pope, on Shroud of Turin, 188–89
Clever Hans, 278–79
Clifford, Edith, 87
Cline, Charlie, 94
coincidences, explanations of, 229–30
cold reading
  meaning of, 145
  mediums, 179, 286
Coleman, Loren, 98
Committee for the Scientific Investigation of Claims of the Paranormal (CSICOP), xiv, 253
Communion: A True Story (Strieber), 126
Communion Letters, The (Strieber), 218, 223
Communion, Transformation, and Breakthrough (Strieber), 224
complementary medicine, New Age fads, 215–16
compostum (cemetery), 107
Comtesse, Marie, 152
Coney Island, side show attractions, 84, 86
confabulation, meaning of, 235
conjoined twins, carnivals, 85, 86
consciousness, altered
  and apparitions, 233–34
  and demonic possession, 15–16, 26–27
  medication-induced, 159–60
  and visions, 111–12
  See also fantasy-prone personality; hypnopompic/hypnagogic hallucinations; lucid dreaming; sleep paralysis
contagion, meaning of, 226
contortionists, carnivals, 87
Coons, Adam, 130
Corbin, Myrtle, 84
Cowdery, Oliver, 302
criminal investigations
  psychic sleuths, 252–56
  remote-viewing case, 61, 65–71

Cromarty, Barbara, 74–75
Cromarty, James, 74–75
crop circles, 115–22
  believers theories about, 166
  crop stamping theory, 120
  hoax, features related to, 116–18, 339
  photos, 119
  vortex effects, 116, 121–22
Crossing Over, 176, 182
  See also Edward, John
crucified persons, carnivals, 86
crying statues. *See* effigies
cryptids, 289–94
  bunyip, 289
  thylacine, 289–91
  yowie, 291–94
curios, of carnivals, 90–91
Cuzco, Peru, puma shape, 5
Cyr, Louis, 87

Damballah (snake *loa*), 141, 144
dancing, and voodoo ceremonies, 143–45
D'Arcis, Pierre, 188–89
Dateline NBC, 179–82, 253, 286
Davenport, Ira, 20, 176–77, 331
  grave of, 332
Davenport, William, 20, 176–77
  Rookwood Cemetery grave, 331–34
Davis, Andrew Jackson, 207
Davis, Richard, 305–6
dead
  communication with, believers, growth of, 31–32
  contacting with Ouija board, 21
  moment-of-death apparitions, 233–35
  See also ghosts; mediums; séances; visitations, of dead
death
  animal knowledge of, 281
  deathbed visions, 235–38
  near-death experience, 223, 238–39
Defense Intelligence Agency (DIA), on Soviet psi research, 62
DeFeo, Dawn, 76–77
DeFeo, Geraldine, 76
DeFeo, Louise, 75
DeFeo, Ronald, Jr., 73, 75, 76–77
DeFeo, Ronald, Sr., 75
Delgado, Pat, 118
Dellafiora, Angela, as remote-viewer, 65–70
demonic possession
  debunking, 15–16, 18–20, 22–23, 25–27
  Exorcist, The, basis of, 14, 17–27
  manifestations of, 14–15, 18–25
Dennett, Michael R., 253
Dennis, John V., 168, 170
dermo-optical perception, 201
detectives, psychic. *See* psychic sleuths; remote-viewing
Devil's Coachhouse, 293
Dickens, Charles, 304, 306
Dickie the Penguin Boy, 85
Diego, Juan, 51, 53
dime museums, 79
ding, 88
Dixon, Jeane, 251
DNA analysis
  alien hybrid, 49
  Shroud of Turin, 194
DNA of God, The (Garza-Valdez), 194
Doolagahl. *See* yowie
dowsing, meaning of, 300
Doyle, Arthur Conan, 58
drawings, unknown sources
  Image of Guadalupe, 51–54
  Shroud of Turin, 192–94
  spirit card writing, 41–42
  spirit precipitations on silk, 34–35
Drbal, Karel, 202
dreams
  about deceased persons, 231–33

dreams *(cont.)*
dream-visions. *See* hypnopompic/hypnagogic hallucinations; lucid dreaming; sleep paralysis
Dreamtime, 291
Dr. Judith Orloff's Guide to Intuitive Healing (Orloff), 215
Dubois, Carl, 173
Durks, William, 86
Dyles, David, 98

Earle, Jack, 83
ectoplasm, 34, 311
Eddy, Mary Baker
as fantasy-prone person, 297
at Winchester Mystery House, 134
Edessa, image of, 195
Edward, John, 176–85
cold readings, 179, 286
discrediting, 180–84
editing of shows, 185
errors, methods of covering, 183
fortuneteller stint, 286
readings, elements of, 178–79, 181–83
Edwards, Frank, 11–12, 56
effigies
Catholic, weeping/bleeding, 324–26
Russian weeping icons, 324–29
El Chupacabras (the goatsucker), 28–30
debunking, 29–30
as predator, 28
El Cid (knight), 242
El Hoppo the Living Frog Boy, 78, 80, 84
embalming, with myrrh, 326
Encyclopedia of Ghosts and Spirits, 135, 157
England
crop circles, 115–22
spontaneous human combustion (SHC) case, 162–64
epilepsy, 15
ESP, Seers & Psychics (Milbourne), 70
Eunus, 59
exorcism
debunking, 18–20, 25–27
demonic possession, manifestations of, 18–25
Exorcist, The (film), real life basis, 14, 17–27
extinct animals. *See* cryptids
extrasensory perception (ESP)
and scientific research, 64
See also clairvoyance; mediums; pet psychics; psychic pets; psychic sleuths; remote-viewing
extraterrestrials
alien hybrid, 46–49
crop circles, 115–22
kidnapping humans. *See* UFO abductions
Nazca lines, 1–9
psychological rationale, 171

faith-healing. *See* healers and healing
fakirs
carnivals, 87
Subbayah Pullavar, 338
Falk, Ruma, 230
fantasy-prone personality
Antoinette Matteson, 210
Benny Hinn, 269
famous cases, 297
Joseph Smith, 296–303
Shirley MacLaine, 137
Syliva Brown, 137
traits of, 111, 210, 296–97, 303
UFO abductees, 224
Farley, James, 305–7
Fate, 11
Feejee Mermaid, 79, 90
fertility rituals, voodoo, 144
festivals
Fisher's ghost, 305
at Kolomenskoye, 258–60
voodoo, 145

fetus, of alien hybrid, 46–49
Finley, Rush, 94
Finucane, R.C., 313
Fiore, Edith, 227
fire
  chimney effect, 163–64
  spontaneous human combustion (SHC), 10–13
  wick effect, 163
fire breathing
  carnivals, 87
  Houdini on, 58, 60
  human blowtorch, 56–60
  method of trickery, 58–60
fire eaters, carnivals, 87
Fischer, John F., 34, 53, 116, 118, 194
Fisher, Frederick, 304–9
  debunking, 306–9
  sightings, 305–6
"Fisher's Ghost" (Dickens), 304, 306
Fitzpatrick, Sonya, 282–84
Flatwoods Monster, 98
Florida, and *El Chupacabras* (the goatsucker), 30
folktales
  "booger" tales, 171–72
  supernatural legends, 308–9
Fonda, Claudia, 279
footprints. *See* tracks/footprints
Ford, Harlan E., 166, 168–69, 173, 174
Ford, Perry, 171
Fort, Charles, 56, 60, 335–40
  science, challenges to, 335–37
fortune-telling
  astrology, 248–51
  See also clairvoyance
Fouke Monster, 170
Four-Legged Girl From Texas, 84
48 Hours, 37
Fowler, Lorenzo, 316
Fowler, Orson Squire
  areas of expertise of, 316–17
  and octagon houses, 316–19, 322–23
Fox, Katie, 20, 31
Fox, Margaret, 20, 31
France, Shroud of Turin, 187–90
freak shows. *See* carnivals
Frei-Sulzer, Max, 194–95, 245–46
French, Chris, 69
French Quarter, New Orleans, 149–50
Friderici, Gottlieb, 46–47
Frog Boy, 84. *See also* El Hoppo the Living Frog Boy
Fuller, Buckminster, geodesic dome, 322
Full Gospel Tabernacle, 262

Gaddis, Vincent, 56, 58
gaffed (faked) freaks, carnivals, 86
Gardner, Martin, 212, 338
Garza-Valdez, Leoncio, 194, 197
Gatorman, 170
Gef (talking mongoose), 280
Geller, Uri, 338
genetic testing, of alien hybrid, 49
geodesic dome, 322
Germany, alien hybrid, 46–49
Ghost Guide to Australia (Davis), 305
ghosts
  Fisher's ghost, 304–9
  Hyde Park Barracks haunting, 271–76
  of Jenolan Caves House, 293
  Marie Laveau sightings, 147, 156–60
  of Moscow, 311–14
  in photographs, 231
  of Santiago de Compostela, Cathedral of, 112
  sightings, explaining, 112, 137–38, 273–74, 313
  of Winchester Mystery House, 135–38
  See also visitations, of dead
Giant Rat, 89–90
giants, carnivals, 83, 86, 90–91
Gilroy, Rex, 291–92, 293–94
Girl in the Fishbowl, 88
Girl with Four Legs and Three Arms, 84
Glapion, Christophe de, 142

glossolalia
meaning of, 16, 265
See also languages, strange
goatsucker. *See El Chupacabras* (the goatsucker)
Goldman, Jane, 56
Golod, Alexander, 203
Goodman, Linda, 248, 250–51
Good Morning, Holy Spirit (Hinn), 262
Gould, Rupert T., 290
Graff, Dale E., 67, 69–70
Great Omi, The Zebra Man, 86
Great Serpent Mound, 5
Green, Bill, 65, 68
Gresham, William Lindsay, 80, 83
grisaille (monochromatic) technique, 193–94
gris-gris (magic charms), 143, 146–47
Guadalupe, image of Virgin Mary, 51–54
as human painting, 52–54
Guscin, Mark, 242

Haiti, and voodoo, 141
Hall, Mark A., 98
Halloran, Father Walter, 17, 23, 24–25
hallucinations
and near-death experience, 238
See also consciousness, altered states; hypnopompic/hypnagogic hallucinations; sleep paralysis
Hall, Ward, 91
Hamad, Mjaka, 124–25
Hamel-Noble, Karen, 285
Haney, Sid, 7
Harris, Martin, 301
Hauck, Dennis William, 105, 130, 158, 311
Haunted City (Dickinson), 156–57
haunted houses
Hyde Park Barracks, 271–76
Hydro Majestic Hotel, 292
investigating, 135–38
Jenolan Caves House, 293
octagonal, Hydes' House, 319–20
Winchester Mystery House, 128–38
Haunted Places: The National Directory (Hauck), 158
Headless Woman, 88, 91
healers and healing
of animals, 285
Benny Hinn, 261–69
dangers of healers, 268–69
and fantasy-prone persons, 210, 269, 297, 303
Joseph Smith, 302
Marie Laveau II, 148
medical intuitives, 207–16
naturalistic reasons for, 111, 148, 267–68
New Age fads, 215–16
Santiago de Compostela cathedral relics, 111
shotgun technique, 262–63
Healing: A Doctor in Search of a Miracle (Nolen), 267
Heller, John, 194
hematite, apport trick, 44
Henry, Asa, 96
Herron, Alan, 30
Hess, Bernard J., 11–13
Heth, Joice, 79
Hidden Memories: Voice and Visions from Within (Baker), 15
Higgins, William, 66
Highland Fat Boys, 79
Hill Cumorah, 301
Hinn, Benny, 261–69
biographical information, 261–62
criticism by Christian groups, 262
healing, explanations of, 266–68
performance of, 263–66
"shotgun" technique, 262–63
hirsute people, carnivals, 84
Hockenberry, John, 181–82
Holyfield, Dana, 166, 168
Holy Shroud Guild, 193

Home for All, A (Fowler), 317, 322
homeopathy, basis of, 319
Honey Island Swamp Monster, 165–74
  description, 166
  discrediting, 166–74
  tracks of, 165, 168–70
Hopkins, Budd, 226
Hornby, Edmund, 234–35
hot-air balloons, Nazca lines theory, 2–3
Houdini, Harry, 20
  and Arthur Conan Doyle, 58
  on breathing fire, 58, 60
  mediums, exposing as fakes, 32, 177, 184, 185, 332
  visit to William Davenport's grave, 332–33
  at Winchester House, 134
Hough, Joseph C., 269
Hough, Peter, 237
House of Blood, 337–38
Household Words, 304
Huey the Pretzel Man, 87
Hughes, Father E. Albert, 22
Human Blockhead, 87
human blowtorch, 56–60
  methods of trickery, 58–60
Human Cigarette Factory, 84
Human Claw-Hammer, 87
human oddities, of carnivals, 79, 80–86
Human Pincushion, 87
hybridization, alien hybrid, 46–49
Hyde, Erastus C., octagon house of, 319–22
Hyde, Julia E., octagon house of, 319–22
Hyde Park Barracks haunting, 271–76
  explaining, 274–76
  sightings, 273–75
Hydes' House
  octagon house, 319–20
  photos, 318
Hydro Majestic Hotel, 292
hydrotherapy, 316
Hyman, Ray, 64
hypnopompic/hypnagogic hallucinations
  features of, 125, 219–23
  of Joseph Smith, 300, 303
  and near-death experience, 223, 238
  and out-of-body experience, 125, 221–22
  and sexual manifestations, 125–26
  and UFO abductions, 126, 219–23, 225
  visitations from deceased in dreams, 232–33, 285
  voices, hearing, 222
  Zanzibar demons, 125–26
  See also sleep paralysis
hypnosis
  stage-hypnotists, choosing subjects, 264–65
  suggestiblity to, 264–65, 296
hysteria, 15

ichthyosis, carnival alligator people, 84, 86
icons, weeping, 324–29
  of myrrh, 326–29
  Nicholas, II, Russian czar, 327–29
illusion show, carnivals, 87–88
Image of Edessa, 195
imaginary friends, 211, 225–26, 296
incubus, 125
indulgences, Catholic, 103
In Search of, 166
Inside Edition, 182
International Circus Sideshow Museum & Gallery, 91
International Directory of Haunted Places, The (Hauck), 105, 130, 311
International Explorers Society, 2
Iona, shroud of Christ, 188
Isbell, William H., 4
Italy
  Shroud of Turin, 187–98
  weeping effigy, 325

Ivan the Terrible, Russian czar, 258
ghost of, 311–12

Jack O'Lantern, 172
Jacobs, David, 222, 226
Jacobson, Susan, 256
Jahoda, Gustav, 230
Jaroff, Leon, 180
Jeanne des Anges, stigmata of, 15
Jefferson, Thomas, octagon house of, 315–16
Jeffries, Anne, 126
Jenolan Caves, 292–94
Jenolan Caves House, 293
Jerry Springer Show, The, 277, 282
Jesus
image of. *See* Shroud of Turin
Oviedo Cloth, 241–46
relics of, 187–88
Joan of Arc, 297
John, Doctor, voodoo doctor, 142–43
John Paul II
on demonic possession, 16
Juan Diego beatification, 53, 54
Jordan, Charlie, remote viewing case, 61, 65–71
Jordan, Otis, 84
Jordan, Peter, 75
juju. See gris-gris
Jung, Carl, 229

karma, meaning of, 104
Kathryn Kuhlman Foundation, 262
Katie, Psychic, 337
Keene, M. Lamar, 34, 35–36, 38, 40, 41, 43–44, 180
Kelley, Ned, 275
Kirlian photography, aura, scientific nature of, 201
Kolomenskoye stone, 258–60
Kosok, Paul, 2
Krantz, Grover, 169, 170
Kruglyakov, Edward, 200, 205
Kuhlman, Kathryn, 262, 268
Kurtz, Paul, 112
Kushner, Harold S., 269
Kuvakin, Valeriĭ, 200, 203–4, 327, 329
Kuznetsov, Dmitrii, 196–97

Lady Wonder (telepathic horse), 279
Lang, John, 306
Langrast, J.B., 146–47
languages, strange
and demonic possession, 16, 22, 26
explaining, 26
speaking in tongues, 265–66
Laveau, Marie, 142–60
biographical information, 142–43
as fortune-teller, 145–46
Marie II (daughter), 148
political-social connections, 143, 145–46
sightings after death, 147, 156–60
tomb of, 152–56
voodoo activities of, 142–45
Laveau, Marie II (daughter)
as healer, 148
tomb of, 153–54
Laveaux, Charles, 142
Layne, Al, 211–12
Learned Pig, 278
Legend of Boggy Creek, The (film), 172
legends, creation of, 107–8
Le Grande Zombi, 144
Leigh, Gerri, 282–83
Lemuria, 104
Lenin, V.I., ghost of, 312–13
Leona the Leopard Girl, 84
Leonard, Gladys Osborne, 297
Le Roy, William, 87
Letiche, 172
Levengood, W.C., 120
levitation
explaining, 43, 338
of spirit trumpets, 43
Subbayah Pullavar, 338

ley lines, meaning of, 104, 112
Libbera, Jean, 84
Life Transitions Center, 228
light orbs, of crop circles, 121–22
Lily Dale (New York), octagon houses, 320–23
Lily Dale Assembly, 32
Lily Dale Mediums' League, 322
Lionel the Lion-faced Man, 84
loa (spiritual entities), 141, 144
Loftus, Elizabeth, 160
Looland, George, 308
Louisiana
  New Orleans and voodoo, 140–50
  swamp monsters, 165–74
Louis I, duke of Savoy, Shroud of Turin, 190–91
loup-garou (Cajun werewolf), 145, 172
lucid dreaming
  meaning of, 111, 223, 232
  and physical/mental exhaustion, 111–12, 223
  of Shirley MacLaine, 104–5, 111–12
  and visions, 111
Lutz, George, 73–76
Lutz, Kathy, 73–76

McBrien, Rev. Richard, 26–27
McClintic Wildlife Management Area, 96
McClung, David, 96
McCrone, Walter C., 193–96, 244, 245
McDaniels, Grace, 84
Mackenzie House, xiii
Mack, John, 227
MacLaine, Shirley
  dream-visions of, 104–5, 111–12
  past-life memories of, 104
  pilgrimage to Santiago de Compostela, 103–5, 111
McMunn, Tony, 162
MaGee, John Jr. *See* Edward, John
magnetism (mesmerism), 317
Mahner, Martin, 49
Making Saints (Woodward), 102
man-beasts
  Bigfoot-like creatures, 168, 172–73, 294
  Gatorman, 170
  yowie, 291–93
manimal. *See* man-beasts
Mannix, Dan, 59–60
Man with Three Eyes, 86
Man with Two Bodies, 84
Man with Two Faces, 86
Marie Laveau of New Orleans (Gandolfo), 142–43, 157
Mark Walberg Show, 252
Marrakech, medina entertainers, 79
massage, and healing, 215
Mathis, Jim, 7
Matteson, Antoinette
  as medical intuitive, 208–11
  remedy product of, 208, 211
Matteson, Judah H.R., 208
May, John, 7
Mays, Jerry, 7
maze, Nazca lines theory, 4
Meaden, George Terence, 120
medical intuitives, 207–16
  Antoinette Matteson, 208–11
  Caroline Myss, 214–16
  Edgar Cayce, 207, 211–14
  Judith Orloff, 215
medication-induced hallucinations, 159–60
mediumistic espionage, 145–46, 180
mediums
  apports trick, 42–45
  automatic writing, 66
  billet scam, 38–40
  closed eyes, seeing with, 38–42
  cold readings, 179
  errors, methods of covering, 183, 285
  Houdini exposé of, 32, 177, 184, 185
  information-gathering methods, 145–46, 180–82
  investigating, 180–84

John Edward, 176–85
pet mediums, 285–87
spiritualism, growth of, 32
spirit writing trick, 40–42, 178
trances, 177–78
trumpets for spirit voices, 34, 42–43
See also clairvoyance; séances
memory
and confabulation, 235
recreative aspects, 160
Mendez-Acosta, Mario, 29
Mendez-Acosta, Patricia, 29
mental illness
and demonic possession, 15–16, 26–27
and fakery, 24–25, 338–39
Merbeings, swamp monsters, 169–70
mermaids, carnivals, 79, 88, 90
mesmerism, 317
Mexico
El Chupacabras (the goatsucker), 28–30
Guadalupe, image of Virgin Mary, 51–54
midgets, carnivals, 83, 86
midway, of carnival, 80
Millerites, 317
Mills, Billy, 166, 168–69, 173
Miracle Crusade. *See* Benny Hinn
Miracle-Mongers and Their Methods (Houdini), 58
miracles, at Santiago de Compostela, Cathedral of, 102–3, 111
Miriam, voodoo queen, 149–50
misdirection, in magic, 38
Missing Time (Hopkins), 226
Momo, 170
money-digging, scrying, 300
Monster Midway (Gresham), 80, 83
monsters
belief in, rationale for, 170–71
Boggy Bayou monster, 173–74
discrediting, 166–74
El Chupacabras (the goatsucker), 28–30
Honey Island Swamp Monster, 165–72
human explanations of, 47–48
Mothman, 93–99
Zanzibar demons, 124–26
"Monstrum Humanum Rarissimum" (Friderici), 46–47
Monticello, octagonal room at, 316
Moore, Hugh, 32, 320–21
Moran, Rick, 75
Mormons, Smith, Joseph, 296–303
Moroni, 300–301
Morrison, Tony, 5
Mortado the Human Fountain, 86
Most Diminutive Lady Samson in the World, 87
Mothman, 93–99
investigating, 94–99
local legend, 93–94
sightings, 95
Mothman and Other Curious Encounters (Coleman), 98
Mothman Prophesies, The (Keel), 93
Mule-Faced Woman, 84
Müller, Dr. Dietmar, 49
Mumler, William H., 32
mummies, carnivals, 90
Munito (talking dog), 278
music, produced by ghosts, 134, 136
Muslims, Moorish conquest of Spain, 102
myrrh
for embalming, 326
of weeping icons, 326–29
Myss, Caroline, as medical intuitive, 214–16
Mysterious Lights and Crop Circles (Howe), 121

Napier, John, 170
National Spiritualist Association of Churches, 32

Index

Nauvoo community, 302
Nazca lines, 1–9
  illustrations, 3, 8
  recreation of, 7–9
  theories about, 1–7
Nazca people, 3–4, 6
near-death experience
  features of, 238
  and hypnopompic/hypnagogic hallucinations, 223, 238
  physiological causes, 238
necropolis, Spain, 107, 112
New Age movement
  healing fads, 215
  medical intuitives, 214–16
  pagan ceremonies, 258–60
  pyramid power, 202–5
  sacred sites, and ley lines, 104, 112
  Santiago de Compostela cathedral, 100–112
New Orleans
  Congo Square, 143–45
  voodoo, 140–50
  voodoo tours, 149–50, 154–55
  See also Laveau, Marie
New York
  burned-over district, 317, 322
  octagon houses, 317–23
Nicholas, II, Russian czar, weeping icon of, 327–29
Nickell, Con, 7
Nickell, J. Wendell, 7, 119, 318, 332–33
Night the DeFeos Died: Reinvestigating the Amityville Murders, The (Osuna), 73, 76
Nixon, Bob, 290
Nolen, William A., 267, 268
Nott, Julian, 2
nuns, and stigmata, 15

Occam's razor, 98–99
Occult Family Physician and Botanic Guide to Health (Matteson), 209–10
octagon houses, 315–23
  beliefs about, 316
  in colonial era, 315–16
  Fowler's promotion of, 316–19, 322–23
  geodesic dome, 322
  Hydes' House, 319–20
  photos, 318
The Octagon (Washington, D.C.), 322
Old Hag, 125
Old Melbourne Goal, 275
Olson, Eugene E., 235
Oneida Community, 317
O'Neill, Michael, 180
One Last Hug Before I Go: The Mystery and Meaning of Deathbed Visions (Wills-Brandon), 235–36
Operation White Crop, 118
Opsasnik, Mark, 26
orbs. *See* light orbs
Orloff, Judith, as medical intuitive, 215
osteopathy, 211–12
Ostrander, Sheila, 62, 201–2
Osuna, Rick, 73, 76–77
Otis the Frog Boy, 84
Ouija board
  and demonic possession, 21
  movement, cause of, 21
out-of-body experiences
  and fantasy-prone personality, 297
  hypnopompic/hypnagogic hallucinations, 125, 221–22
  near-death experience, 238
  and remote-viewing, 63–71
  and sleep paralysis, 222–23
Oviedo Cloth, 241–46
  blood, testing for, 243–45
  debunking, 244–46
  history of, 242–43
  link to Shroud of Turin, 242, 243, 244–45
  pollen analysis, 245–46
Oviedo Cloth, The (Guscin), 242

Owlman, 98
owls, Mothman sightings, 96–99

pagan ceremonies, Kolomenskoye stone, 258–60
pain, and power of suggestion, 267
paintings
  Oviedo Cloth, 244
  Shroud of Turin as, 192–94
  See also drawings, unknown sources
Panic of 1857, 317, 319
Pardee, Sarah L., 130
Paris, Jacques, 142, 152
Patterson, Roger, 168
pencil reading, 279
Pepper, John, 304
Persinger, Michael A., 268–69
Personal Freedom Outreach, 262
personality traits
  and demonic possession, 15–16, 26–27
  suggestibility to hypnosis, 264–65
  See also fantasy-prone personality
Peru
  Cuzco, layout of, 5
  Nazca lines, 1–9
pet mediums, 285–87
pet psychics, 282–85
  animal healers, 285
  explaining, 282–84
  finding lost animals, 284–85
  séances by, 286–87
  stunts of, 283–84
Pet Psychic, The (TV show), 282
petrified man, 90
pets, psychic. *See* psychic pets
Pfungst, Oskar, 278
photographic images
  aura, 201
  deceased, images in photographs, 231
  Kirlian photography, 201
  Shroud of Turin, 191–92
  spirit photography, 32
phrenology, 316
Pickens, Ray, 172
Pickett, Thomas J., 197
pilgrimage, Santiago de Compostela, 102–5
placebo effect, and healing phenomenon, 111
plasma vortices, and crop circles, 116, 120–21
Point Pleasant (West Virginia), Mothman, 93–99
pollen analysis
  Oviedo Cloth, 245–46
  Shroud of Turin, 194–96
poltergeists
  explaining, 18–20
  manifestations of, 18–21
  and spiritualism, 20
Ponchartrain, Lake, 144
popobawa. *See* Zanzibar demons
Possessed (Allen), 17
possession
  by *loas* in voodoo, 141, 144
  See also demonic possession
precipitations on silk, 34–35
precognitive perception, 63
Primetime, 76
Prince Randian, the Hindu Living Torso, 84
psi research
  of animals, 281–82
  remote-viewing, 61–71
  Soviet Union, 62
Psychic Discoveries Behind the Iron Curtain (Ostrander and Schroeder), 62, 201–2
Psychic Mafia, The (Keene), 38, 44, 180
psychic pets, 279–81
  explaining, 279–82
psychics
  Barnum effect, 283–84
  cold readings, 145, 179, 286
  errors, methods of covering, 183, 285

as fantasy-prone personalities, 111, 137, 210, 297
See also clairvoyance; mediums; pet psychics; psychic pets; psychic sleuths
psychic sleuths, 252–56
discrediting, 254–56
Dorothy Allison, 253–56
retrofitting technique, 70, 255–56
Psychic Sleuths (Dennett), 253
psychokinesis
meaning of, 62
Russian interest, 62, 201
psychokinetic energy, and poltergeist, 18
Psychology of Superstition, The (Jahoda), 230
Puerto Rico, *El Chupacabras* (the goatsucker), 28–30
Pullavar, Subbayah, 338
puma shape, Cuzco, Peru, 5
Puthoff, Harold, 62–63
pyramid power
claims about, 202–3, 205
photos, 204
Russia, 202–5

qi (life force), 201

Radford, Benjamin, 25, 119
radiocarbon dating, Shroud of Turin, 196, 241–42
Rambo, Ralph, 133–34
Ramsay, JonBenet, 253–54
Randi, James, 213–14, 338–39
Randles, Jenny, 237
rape
incubus and *succubus*, 125–26
Zanzibar demons, 124–26
Reeser, Mary, 10, 339–40
Rees, Vaughn, 129, 137
reflexology, 215
Reiche, Maria, 2, 6
reincarnation
and *karma*, 104
Shirley MacLaine, past-lives of, 104
relics of saints
medieval relic-mongering, 109
nature of, 102
Oviedo Cloth, 241–46
Shroud of Turin, 187–98
St. James the Greater, 102–12
religion and supernatural
icons, weeping, 324–29
visionaries, Joseph Smith, 296–303
voodoo, 140–50
See also Roman Catholicism
remote-viewing, 61–71
Charlie Jordan case, 61, 65–71
conclusions about, 69–71
definition of, 63
research methods, 63–64
retrofitting, 70
scientific investigations of, 62–65
Resch, Tina, 338–39
Resurrection of the Shroud, The (Antonacci), 187
retrocognitive perception, 63
retrofitting
meaning of, 70, 255
psychic sleuths, 255–56
Reynolds, Bobby, 89, 91
Rhine, Dr. J.B., 279
Rice, Anne, 140
Rice, Madame, 87
Richeliue, Cardinal, 15
Rich-Twinn Octagon House, 322
Ridler, Horace, 85–86
Riffle, Mable, 35–37
Riley, Mel, 66
Ripley, Robert, 78
Ritter, Archbishop Joseph, 23, 25
rituals
for rain, 4
voodoo, 144–45
River Dreams (Graff), 69
Roads to Santiago (Nooteboom), 105, 107

Robertson, Pat, 262
Rodgers, Peter, 292, 304, 309, 332–33
Roman Catholicism
exorcism, 15–16, 17, 22–27
image of Guadalupe, 51–54
indulgences, 103
Oviedo Cloth, 241–46
relics of saints, 102, 109
Santiago de Compostela, Cathedral of, 101–12
Shroud of Turin, 187–98
and voodoo, 141, 143, 149
weeping/bleeding effigies, 324–26
Rookwood Cemetery, 331–34
William Davenport's grave, 331–34
Roosevelt, Theodore, 134
rubes (outsiders), 91
Russia
ghosts of Moscow, 311–14
icons, weeping, 324–29
Kolomenskoye stone, 258–60
psychic studies/interests of, 200–201
pyramid power craze, 202–5
Russian Orthodox Church, icons, weeping, 324–29

Sainte Chapelle (Paris), 109
saints
James and Santiago de Compostela cathedral, 111–12
Virgin Mary, images of, 51–54
and voodoo, 141
weeping effigies, 184–85, 324–29
Salem witch hysteria, 226
Santiago de Compostela, Cathedral of, 100–112
apparitions of pilgrims, 112
Camino route, 103–5
investigation/explanations, 105–12
miracles, 102–3, 111
photos, 106
pilgrimage to, 102–5
relics at, 102–3, 106–7
remains of St. James the Greater, 101–2
Sardinia, bleeding effigy, 325
sasquatch, 291
carnival exhibit, 90
See Bigfoot-like creatures
Satan
banishing. *See* Exorcism
possession by, 22–27
Sawyer, Alan, 4
Scape Ore Swamp Lizardman, 170
Scarberry, Linda, 95
Schafersman, Steven D., 195
schizophrenia, 15
Schmiedt, Andreas, 46
Schmiedt, Johanna Sophia, 46–48
Schnabel, Jim, 66, 68
Schroeder, Lynn, 62, 201–2
Schulemburg, Guillermo, 51, 53–54
Science Applications International Corporation, remote-viewing research, 62
scrying, meaning of, 300
Sealo the Seal Boy, 84
séances
apports trick, 42–45
believers, growth of, 31–32
for deceased pets, 286–87
in octagon houses, 322–23
spirit voices, production of, 34, 42–43, 177
Search for the Twelve Apostles (McBirnie), 109–10
Second Sight (Orloff), 215
selection fallacy, 230
Senchihina, Uliya, 203–4
Severs, Henrietta, 131
sexual manifestations
incubus and *succubus*, 125–26
voodoo rituals, 144–45
Zanzibar demons, 124–26
Shakers, 317
Shaw, Allen, 332–33
Shealy, C. Norman, 215

Sheldrake, Rupert, 281–82
"shotgun" technique, faith-healing, 262–63
Shroud of Turin, 109, 187–98
 accelerator mass spectrometry (AMS) dating, 196–97
 DNA analysis, 194
 examinations of, 192–98
 in faith-healing scam (1300s), 188–89
 flower and plant images on, 197–98
 history of, 187–92
 link to Oviedo Cloth, 242, 243–44
 paint traces found, 192–94
 photographic images, 191–92
 pollen analysis, 194–96
 radiocarbon dating, 196, 241–42
 Vatican possession of, 190
Shroud of Turin Research Project (STURP), 193–94
Siamese twins
 Adolph and Rudolph, 86
 Chang and Eng, 85
sideshows. *See* carnivals
signs
 coincidences, 229–30
 from deceased, 220–31
Signs (film), 115
Silverman, Helaine, 4
simulacra, 231
Sisman, Alan, 335–40
Skeptical Inquirer, xiv, 25, 197, 338, 339
Skretny, Judith, 228
Skunk Ape, 170
slaves, American, and voodoo, 141
sleep-associated phenomena. *See* hypnopompic/hypnagogic hallucinations; lucid dreams; sleep paralysis
Sleeping Prophet. *See* Cayce, Edgar
sleep paralysis
 and hypnopompic/hypnagogic hallucinations, 219, 220–21, 225
 and UFO abductees, 219, 220–21, 225
 visitation from the dead during, 233, 274–75
Smith, Emma, 301
Smith, Hyrum, 302
Smith, Joseph, 296–303
 biographical information, 298–302
 as fantasy-prone personality, 297–98, 301, 303
 money-digging, 300–301
 visions of, 298–300, 302
 writings of, 301–2
Smith, Suzy, 130
Smurl, Jack, 125–26
snakes
 charmers, carnivals, 81, 87
 and voodoo, 141, 144
Sol Rosales, José, 54
Soviet Union
 collapse of, 200
 psi research, 62
 See also Russia
Spain
 Oviedo Cloth, 241–46
 Santiago de Compostela, Cathedral of, 101–12
speaking in tongues, 265–66
Spidora the Spider Girl, 88
"Spirit of the Creek, The" (poem), 307
spirits. *See* ghosts; mediums; séances; visitations, of dead
spiritualism
 Camp Chesterfield enclave, 31–45
 camp meetings, 32–33
 and Davenport brothers, 20, 176–77, 331–34
 exposés, 32–37, 176–77, 320–21
 Lily Dale village, 320–23
 origin and growth of, 20–21, 31–32, 176–77, 285–86
spontaneous human combustion (SHC), 10–13
 cases of, 10–12, 339–40
 explaining, 10, 12–13, 162–64

spontaneous remission, and healing phenomenon, 111, 266
spys and intelligence, remote-viewing, 61–71
stage-hypnotists, subjects, choosing, 264–65
Stalin, Joseph, 312–13
Stanford Research Institute (SRI), remote-viewing research, 62–64
Stargate, remote-viewing activity, 62, 67–68
Steiger, Brad, 235
stelae (tombstones), 107
Step Right Up! (Mannix), 59–60
stigmata, Jeanne des Anges, 15
St. James the Greater
  healing by relics, 111
  life of, 101, 107, 109–11
  remains at Santiago de Compostela, 102–12
St. John's Eve, voodoo ritual, 145
St. Patrick, snake symbol, 141
Stranger Than Science (Edwards), 11
Stratton, Charles Sherwood (Tom Thumb), 82, 83
street performers, 78–79
Strieber, Whitley, 126, 218–26
strong people, carnivals, 87
succubus, 125
sudarium (napkin), 241
  See also Oviedo Cloth
suggestion, power of
  and contagion, 226
  and ghost sightings, 313
  and healing phenomenon, 111, 148, 267
  and hypnosis, 264–65, 296
Sullivan, Helen, 268
Sun Signs (Goodman), 250
Superminds (Taylor), 338
supernatural legends, 308–9
suspensions, and levitation, 338
swamp monsters
  Bigfoot-like, 168, 172–73
  bunyip, 289
  Honey Island Swamp Monster, 165–72
sword swallowers, carnivals, 87
synchronicity, meaning of, 229

talker, at carnival, 80
talking animals, 277–80
  history of, 277–80
Tallant, Robert, 143, 149, 153, 158
Targ, Russell, 62–63
Tasmanian tiger. *See* thylacine
tattooed persons, carnivals, 85–86
Taylor, John, 338
telepathy. *See* pet psychics; psychic pets; psychic sleuths; remote viewing
telesthesia
  meaning of, 63
  remote-viewing, 63–71
temperance movement, 317
ten-in-one shows, 80–83
Tereshina, Nadya, 258, 259
Texas Giant, 83
That's Incredible, 75
Theisen, Donna, 229, 232
Theosophy, founder, 297
therapeutic touch, 215
They Call It Hypnosis (Baker), 264
Threat, The (Jacobs), 226
Thummin, 300
thylacine, 289–91
  extinction of, 289–90
  photo, 290
  sightings, 290–91
Toby the Sapient Pig, 278
Tocci brothers, two-headed boy, 83
tombs
  of Marie Laveau, 152–56
  William Davenport's grave, 331–34
Tom Thumb, wedding of, 82, 83
Toronto, oil-weeping effigy, 325
torture acts, carnivals, 87

Tourette's syndrome, 15
tracer powders, 18–19
tracks/footprints
  Bigfoot-like, 168, 172–73
  discrediting, 172–73
  Honey Island Swamp Monster, 165, 168–70
  yowie, 293–94
Tracks in the Psychic Wilderness (Graff), 67
trances
  and Edgar Cayce, 212
  mediums, 177–78
Trinovantes, 108
trumpets
  apports from, 43–44
  levitation of, 43
  spirit voices from, 34, 42–43, 177
Two Curious Birds, 278
Two-Headed Boy, 83

UFO abductions, 218–27
  accounts of, 219–26
  waking-dream theory, 126, 219–23
Underwood, William, 57–60
Unsolved Mysteries, 337
Urim, 300
utopianism, 317
Utts, Jessica, 64

Van Praagh, James, 177
  debunking, 286
vaporography, 192
Vargas, Elizabeth, 76
Vatican, on exorcism, 15, 16
Virgin Mary
  Image of Guadalupe, 51–54
  weeping icons, 326–27
visions. *See* hypnopompic/hypnagogic hallucinations; lucid dreaming
visitations, of dead, 228–39
  dead pets, 285–87
  deathbed visions, 235–38
  dream contacts, 231–32
  explaining, 223, 229–30, 232–33, 238, 285
  moment-of-death apparitions, 233–35
  signs, 220–31
  See also ghosts
vitiligo, carnival leopard girl, 84
vodun (spirit), 141. *See* voodoo
voices
  hypnopompic/hypnagogic hallucinations, 222
  spirits, at séance, 34, 42–43, 177
Von Däniken, Erich, 1–2
voodoo, 140–50
  altars, 150
  historical view, 140–41
  modern practice, 149–50
  photos, 144, 150
  queen. *See* Laveau, Marie
  rituals, 144–45
  Roman Catholic elements of, 141, 143, 149
  voodoo doctors, 142–43
Voodoo in New Orleans (Tallant), 143
Voodoo Spiritual Temple, 149–50
vortex effects, crop circles, 116, 121–22

Wagner, Dr. Paul, 167
Wagner, Sue, 167
waking dreams
  basis of, 125
  See also hypnopompic/hypnagogic hallucinations; sleep paralysis
Warner, Steve, 96
Warren, Ed, 125–26
Warren, Lavinia, 83
Washington, George, octagon house of, 315–16
Water Cure Journal, 316
Watson, Lyall, 335–36
Weber, William, 75
weeping icons. *See* icons, weeping
weight-lifters, carnivals, 87

werewolf, Cajun, 145, 172
Whanger, Alan, 197, 244–45
Whanger, Mary, 197
Whitcomb, Nellie, 211
Whitman, Walt, 316
Whitstine, Earl, 173
Why People Don't Heal and How They Can (Myss), 214–15
wick effect, 163
Williams, Betty Lou, 84
Wills-Brandon, Carla, 235–36
Wilson, Ian, 194, 198, 244
Wilson, Sheryl C., 296–97
Winchester, Annie, 130
Winchester Mystery House, 128–38
  facts versus fantasy about, 130–38
  features of, 131–33
  haunting, explaining, 135–38
  photos, 129, 132, 137
Winchester, Oliver Fisher, 130
Winchester, Sarah, 128–38
Winchester, William Wirt, 130
wolf children, 337
Wonderful Intelligent Goose, 278
wonder-workers, of carnivals, 87
Woodman, Dr. L.C., 56–58
Woodman, Jim, 2
Woodward, Kenneth L., 102
World's Most Incredible Stories: The Best of Fortean Times (Sisman), 335–40
Worrell, George, 305
writing, unknown sources
  and demonic possession, 15, 22–25
  slate trick, 40
  spirit card writing, 40–42, 178

yeti, 291
yowie, 291–94
  questioning, 293–94
  sightings, 291–92

Zanzibar demons, 124–26
  waking dream theory, 125–26
Zaragoza, Patricia Lopez, 53–54
zombies, 172